Modern Methods of Plant Analysis

Volume 18

Editors
H.F. Linskens, Nijmegen/Siena/Amherst
J.F. Jackson, Adelaide

Volumes Already Published in this Series:

Volume 1: Cell Components
1985, ISBN 3-540-15822-7

Volume 2: Nuclear Magnetic Resonance
1986, ISBN-3-540-15910-X

Volume 3: Gas Chromatography/Mass Spectrometry
1986, ISBN 3-540-15911-8

Volume 4: Immunology in Plant Sciences
1986, ISBN 3-540-16842-7

Volume 5: High Performance Liquid Chromatography in Plant Sciences
1987, ISBN 3-540-17243-2

Volume 6: Wine Analysis
1988, ISBN 3-540-18819-3

Volume 7: Beer Analysis
1988, ISBN 3-540-18308-6

Volume 8: Analysis of Nonalcoholic Beverages
1988, ISBN 3-540-18820-7

Volume 9: Gases in Plant and Microbial Cells
1989, ISBN 3-540-18821-5

Volume 10: Plant Fibers
1989, ISBN 3-540-18822-3

Volume 11: Physical Methods in Plant Sciences
1990, ISBN 3-540-50332-3

Volume 12: Essential Oils and Waxes
1991, ISBN 3-540-51915-7

Volume 13: Plant Toxin Analysis
1992, ISBN 3-540-52328-6

Volume 14: Seed Analysis
1992, ISBN 3-540-52737-0

Volume 15: Alkaloids
1994, ISBN 3-540-52738-9

Volume 16: Vegetables and Vegetable Products
1994, ISBN 3-540-55843-8

Volume 18: Fruit Analysis
1995, ISBN 3-540-59118-4

Forthcoming:
Volume 17: Plant Cell Wall Analysis
1996, ISBN 3-540-59406-X

Fruit Analysis

Edited by
H.F. Linskens and J.F. Jackson

Contributors

M.S. Allen M.A. Berhow H.S.M. de Vries C.H. Fong
S. Hasegawa J.F. Jackson N. Kubota H.F. Linskens
G. Mulcahy Bergamini D.L. Mulcahy E. Pesis S. Taira
A. Trautweiler A. Vezvaei P.J. Williams J.M. Witherspoon

With 36 Figures

Springer

Professor Dr. HANS FERDINAND LINSKENS
Goldberglein 7
D-91056 Erlangen, Germany

Professor Dr. JOHN F. JACKSON
Department of Horticulture, Viticulture, and Oenology
Waite Agricultural Research Institute
University of Adelaide
Glen Osmond, S.A. 5064
Australia

ISBN-13:978-3-642-79662-3 e-ISBN-13:978-3-642-79660-9
DOI: 10.1007/978-3-642-79660-9

The Library of Congress Card Number 87-659239 (ISSN 0077-0183)

Softcover reprint of the hardcover 1st edition 1995

Typesetting: Best-set Typesetter Ltd., Hong Kong
SPIN 10134144 31/3137 - 5 4 3 2 1 0 - Printed on acid-free paper

Introduction

Modern Methods of Plant Analysis

When the handbook *Modern Methods of Plant Analysis*, was first introduced in 1954, the considerations were:

1. the dependence of scientific progress in biology on the improvement of existing and the introduction of new methods;
2. the difficulty in finding many new analytical methods in specialized journals which are normally not accessible to experimental plant biologists;
3. the fact that in the methods sections of papers the description of methods is frequently so compact, or even sometimes to incomplete, that it is difficult to reproduce experiments.

These considerations still stand today.

The series was highly successful, seven volumes appearing between 1956 and 1964. Since there is still today a demand for the old series, the publisher has decided to resume publication of *Modern Methods of Plant Analysis*. It is hoped that the New Series will be just as acceptable to those working in plant sciences and related fields as the early volumes undoubtedly were. It is difficult to single out the major reasons for the success of any publication, but we believe that the methods published in the first series were up-to-date at the time and presented in a way that made description, as applied to plant material, complete in itself with little need to consult other publications.

Contribution authors have attempted to follow these guidelines in this New Series of volumes.

Editorial

The earlier series of *Modern Methods of Plant Analysis* was initiated by Michel V. Tracey, at that time in Rothamsted, later in Sydney, and by the late Karl Paech (1910–1955), at that time at Tübingen. The New Series will be edited by Paech's successor H.F. Linskens (Nijmegen, The Netherlands) and John F. Jackson (Adelaide, South Australia). As were the earlier editors, we are convinced "that there is a real need for a collection of reliable up-to-date methods for plant analysis in large areas of applied biology ranging from agriculture and horticultural experiment stations to pharmaceutical and technical institutes concerned with raw material of plant origin".

The recent developments in the fields of plant biotechnology and genetic engineering make it even more important for workers in the plant sciences to become acquainted with the more sophisticated methods, which sometimes come from biochemistry and biophysics, but which also have been developed in commercial firms, pharmaceutical laboratories, non-university research institutes, and medical establishments.

Concept of the New Series

Many methods described in the biochemical, biophysical, and medical literature cannot be applied directly to plant material because of the special cell structure, surrounded by a tough cell wall, and the general lack of knowledge of the specific behaviour of plant raw material during extraction procedures. Therefore all authors of this New Series have been chosen because of their special experience with handling plant material, resulting in the adaptation of methods to problems of plant metabolism.

Nevertheless, each particular material from a plant species may require some modification of the described methods and usual techniques. The methods are described critically, with hints as to their limitations. In general it will be possible to adapt the methods described to the specific needs of the users of this series, however references have been made to the original papers and authors. During the planning of this New Series the editors tried to ensure that the aims and general layout of the contributions are within the general guidelines indicated above, but in addition they tried not to interfere too much with the personal style of each author.

There are several ways of classifying the methods used in modern plant analysis. The first is according to the technological and instrumental progress made over recent years. These aspects were taken into consideration in Volumes 1 to 5 and 11 of this series describing methods in a systematic way according to the basic principles of the methods.

A second classification is according to the plant material that has to undergo analysis. The specific application of the analytical method is determined by the special anatomical, physiological, and biochemical properties of the raw material and the technology used in processing. This classification was used in Volumes 6 to 8, 10, 14, 16 and 18.

A third category is according to the classes of substances present in the plant material and the subject of analytical methods. The latter was used for Volumes 9, 12, 13 and 15 of the series.

Naturally, these three approaches to developments in analytical techniques for plant materials cannot exclude some small overlap and repetition; but careful selection of the authors of individual chapters, according to their expertise and experience with the specific methodological technique, the group of substances to be analyzed, or the plant material which is the subject of chemical and physical analysis, guarantees that recent developments in analytical methodology are described in an optimal way.

Volume Eighteen – Fruit Analysis

This volume on *Fruit Analysis* begins with general chapters on non-intrusive fruit analysis and post-harvest treatments of fruit which maximize fruit aroma and quality. During fruit ripening, some changes are degradative (e.g. chlorophyll breakdown, starch hydrolysis and cell wall degradation) and others synthetic (e.g. synthesis of aroma volatiles, carotenoids and anthocyanins). In order to enhance the aroma volatiles which contribute to fruit flavour, taste and smell, all of the above processes need to be understood and taken into account to be able to provide suitable post-harvest treatments. The second chapter deals with measurements of sensory quality and then outlines the exogenous application of metabolites as an enhancement procedure, followed by the application of anaerobic conditions for the induction of aroma volatiles. Methods for analysis of flavouring compounds in table grapes are dealt with in the third contribution. While there have been many descriptions of methods for analyzing the varietal flavour of wine grapes, the specific flavour attributes of table grapes have attracted little attention, and so this chapter is timely. It begins with methoxypyrazine analysis, followed by glycosidically bound flavour compounds of grapes and methods used in their analysis, either as intact glycoside precursors (gas chromatography, HPLC, countercurrent chromatographic methods, mass spectrometry) or after hydrolysis of the glycosides. The work by these authors breaks new ground and therefore fills a very important place in this volume. Turning next to citrus fruits, the next contribution deals with the analysis of bitter principles in citrus. Bitterness is a major problem in the citrus industry, as bitter juices have a lower market value. The problem is caused by the accumulation of limonin from the limonoid terpine group or naringin from the flavonoid phenolic group, both of which may be present in considerable proportions in juice obtained from unripe fruit. Methods for the analysis of these compounds are presented and suggestions for reducing amounts of the bitter components put forward.

Astringency can be a problem in some fruit, and so we have two chapters dealing with this problem. The first takes up the problem of phenolics in peach, and puts forward an analytical approach to the concentration of phenolics and its relationship to phenylalanine ammonia-lyase activity and to astringency in peach fruit. The second chapter on astringency considers the problem in persimmon. In certain persimmon cultivars, the fruit is always astringent. Therefore, they are not commercially important. The so-called sweet persimmons are non-astringent at maturity and are cultivated in many countries. The astringency in persimmons is due to soluble tannins that accumulate in large, specialized cells (tannin cells). In this chapter, an outline of simple, reliable methods for the estimation of astringency and tannins in persimmon is presented.

The fruit of apricot is consumed fresh, or it is juiced or canned, and can also be dehydrated (dried). Many analytical procedures are used to monitor the fruit in all of these processes, including sugar content, acidity and apricot aroma. A collection of methods is presented in the following chapter, which deals with all the practical analytical methods needed by scientists working in the apricot industry.

The volume is rounded off by two chapters dealing with two different aspects of nuts. The first describes a series of analytical methods which are concerned with nutritional aspects of almond nuts. Thus, methods are presented for the determination of oils, proteins, sugars and fibre content in almond nuts, as well as soluble sugars, minerals, gums, mucilage, pectin and cyanide (used when dealing with bitter almond for prussic acid manufacture). Because almond hulls can be used for animal feed, the analysis of hulls is also considered. Finally, the volume concludes with a chapter on the genetic characterization of chestnut. Since so much of today's efforts go into the breeding of superior fruits and nuts, it was thought appropriate here to include an approach making use of random, amplified polymorphic DNA (RAPD) characterization of chestnut varieties to speed up such breeding programmes.

Acknowledgements. The editors would like to express their gratitude to all contributors for their efforts in keeping the production schedules, and to Dr. Dieter Czeschlik and the staff of Springer-Verlag, especially Ms. Isolde Tegtmeier, for their cooperation in preparing this and other volumes of the series, *Modern Methods of Plant Analysis.*

Adelaide and Nijmegen/Siena/Amherst, Spring 1995.

H.F. Linskens
J.F. Jackson

Contents

List of Contributors

ALLEN, M.A., Ron Potter Centre, Charles Sturt University, P.O. Box 588, Wagga Wagga, New South Wales, 2650, Australia

BERHOW, M.A., National Center for Agricultural Utilization Research, United States Department of Agriculture, Agricultural Research Service, 1815 N. University Street, Peoria, Illinois 61604, USA

DE VRIES, H.S.M., Department of Molecular and Laserphysics, University of Nijmegen, Toernooiveld, NL-6525 ED Nijmegen, The Netherlands

FONG, CHI H., 527 San Pascual Street, Los Angeles, Calfornia 90042, USA

HASEGAWA, S., Western Regional Research Center, United States Department of Agriculture, Agricultural Research Service, 800 Buchanan Street, Albany, California 94710, USA

JACKSON, J.F., Department of Horticulture, Viticulture, and Oenology, Waite Agricultural Research Institute, University of Adelaide, Glen Osmond, S.A. 5064, Australia

KUBOTA, N., Faculty of Agriculture, Okayama University, Tsushima, Okayama 700, Japan

LINSKENS, H.F., Goldberglein 7, D-91056 Erlangen, Germany

BERGAMINI MULCAHY, G., Department of Biology, University of Massachusetts, Amherst, Massachusetts 01003 USA

MULCAHY, D.L., Department of Biology, University of Massachusetts, Amherst, Massachusetts 01003 USA

PESIS, E., Department of Postharvest Science of Fresh Produce, Institute for Technology and Storage of Agricultural Products, Agricultural Research Organization, The Volcani Center, P.O. Box 6, Bet Dagan 50250, Israel

TAIRA, S., Laboratory of Pomology, Faculty of Agriculture, Yamagata University, 1–23 Wakaba-cho, Tsuruoka-City, Yamagata-pref. 997, Japan

TRAUTWEILER, A., Department of Biology, University of Massachusetts, Amherst, Massachusetts 01003, USA

Vezvaei, A., Department of Horticulture, Viticulture and Oenology, Waite Agricultural Research Institute, University of Adelaide, Glen Osmond, S.A. 5064, Australia

Williams, P.J., The Australian Wine Research Institute, P.O. Box 197, Glen Osmond, South Australia, 5064, Australia

Witherspoon, J.M., Department of Horticulture, Viticulture and Oenology, Waite Agricultural Research Institute, University of Adelaide, Glen Osmond, S.A. 5064, Australia

Non-Intrusive Fruit and Plant Analysis by Laser Photothermal Measurements of Ethylene Emission

H.S.M. de Vries

1 Introduction

Photothermal trace gas detection reaches extremely high sensitivity combined with good time resolution for a select group of molecules that are of great importance in a biological context. For ethylene, a sensitivity of 6 pl/l has been achieved; for ammonia, 0.5 nl/l has been detected; methane produced by anaerobic bacteria has been measured down to 1 nl/l. The field of application and of investigated gas species is growing rapidly. In this contribution, we focus on ethylene because, for this molecule, this novel technique has been extensively applied.

In general, ethylene acts as a gaseous hormone, involved in and produced during the ripening of (climacteric) fruit (Seymour et al. 1993), abscission of leaves, growth and senescense of plants and the wilting of flowers (Abeles et al. 1992). External stress factors (e.g. chemical, thermal, mechanical, wounding, gamma irradiation) also induce variations in ethylene production (Yang and Hoffman 1984; Kende 1993). As an example of stress, untimely ripening of fruits occurs due to damage during transport.

Ethylene levels are most commonly analyzed by gas chromatographic techniques, first introduced by Huelin and Kennett in 1959. This analytical tool equipped with either a flame ionization detector or a photoionization detector possesses detection limits of 10 and 1 nl/l, respectively (Bassi and Spencer 1989). The development of flow-through systems in line with an ethylene trap, thereafter measured by GC (de Greef and de Proft 1978), allows the exposure of the biological sample to well-defined conditions chosen by the experimenter. For instance, it becomes possible to change the gas composition and to use scrubbers to remove gases either influencing ethylene production or interfering in ethylene detection.

The ethylene trap collects all the produced volatile hormone for a certain period so that small production rates become detectable. Transitory features are thus lost. Very sensitive laser-based ethylene monitors become then advantageous. Using a laser-driven photoacoustic (PA) ethylene detector coupled to a flow-through setup, a detection limit of 6 pl/l is reached with a response time of 4 min at a flow of 1 l/h (Harren et al. 1990a). The photothermal deflection (PTD) instrument monitors ethylene concentrations down to 0.5 nl/l, locally, much faster (1 s) and non-intrusively. In this way, ethylene production, e.g. by fruits, can be measured non-destructively. Very recently, we have improved the time

Modern Methods of Plant Analysis, Vol. 18
Fruit Analysis
Edited by H.-F. Linskens and J.F. Jackson

response for PTD to 0.1 s and for PA to 2 s. It should be noted that PTD experiments can be performed in the open air, thus avoiding wall adsorption problems for molecules with a large dipole moment (e.g. notorious for ammonia; Henningsen 1994) and reactive concentration changes (e.g. observed for ozone). Ethylene, ammonia and ozone possess distinct fingerprints in the CO_2 laser wavelength region (Fig. 1).

PTD and PA are related techniques based on vibrational excitation of molecules using infrared light (Sigrist 1994). The excited molecules transfer their vibrational energy through collisions to translational degrees of freedom, yielding a rise in temperature (Fig. 2). With modulated light, the temperature and, consequently, the pressure change periodically giving rise to the photoacoustic (PA) effect. The modulated pressure change is probed by microphones. PTD also makes use of a locally increased temperature (Jackson et al. 1981). A second light beam traversing nearby is deflected due to the change in the refractive index in the heated region; its displacement is measured with a position-sensitive detector.

In this chapter, attention is mainly paid to technical aspects of the laser methods and to biological applications to illustrate the viability of these methods.

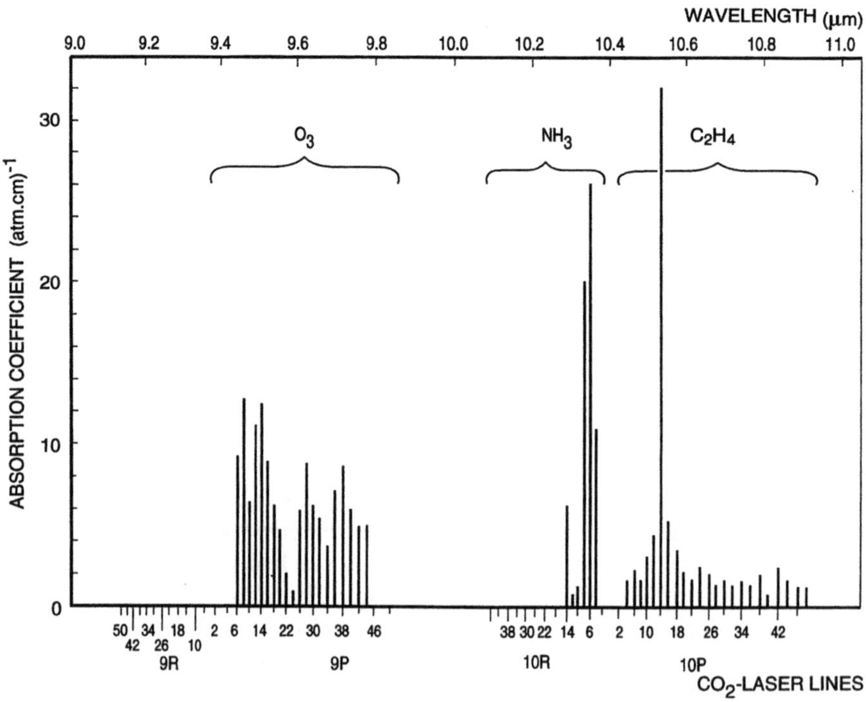

Fig. 1. Main absorption coefficients of ozone (Harren et al. 1990b), ammonia (Brewer and Bruce 1978) and ethylene (Brewer et al. 1982) of the CO_2 laser wavelength region (9–11 μm)

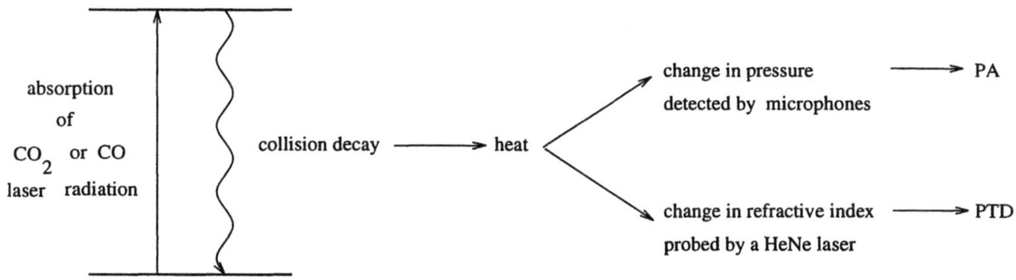

Fig. 2. Two different schemes based on absorption and subsequent collision decay followed by heat production. *PA* Photoacoustics; *PTD* photothermal deflection

Using a CO_2 laser as excitation light source, applications involving ethylene demonstrate the course of events in natural processes like ripening and for species under stress conditions such as wounding, anaerobiosis, ozone exposure and flooding. This chapter ends with a discussion of a photoacoustic setup employing a CO laser instead of a CO_2 laser in sensitive trace gas measurements, e.g. of methane and acetaldehyde.

2 The Photothermal Deflection Method

2.1 The Experimental Setup

The principle of the PTD technique relies on the deflection of a light beam due to a gradient of refractive index. In PTD, the gradient is induced by an intense light beam from a CO_2 laser. This laser is periodically interrupted at 20 Hz by a rotating mechanical chopper (Fig. 3, above). Absorption of this beam by the medium gives rise to a local temperature increase, leading to a temperature gradient with concentric, cylindrical isotherms around the CO_2 laser beam. According to the ideal gas law, this temperature gradient is accompanied by a density gradient and, therefore, by a refractive index gradient. The latter is probed by measuring the deflection of a second light beam from a HeNe laser; the deflection is a direct measure of the amount of CO_2 laser light absorbed. Thus, the deflection of the HeNe laser beam is proportional to the intensity of the CO_2 laser beam and to the density of the absorbing species in the medium, e.g. ethylene molecules in air. Finally, the deflection signal is measured with a position-sensitive detector (PSD; Centronic QD 50-2) from which the signal is fed into a lock-in amplifier.

In the quartz waveguide tube of the CO_2 laser, radiation is generated between 9 and 11 μm at about 90 discrete laser transitions (Fig. 3). The gas mixture inside this tube (inner diameter 3 mm, length 400 mm) is continuously refreshed at a

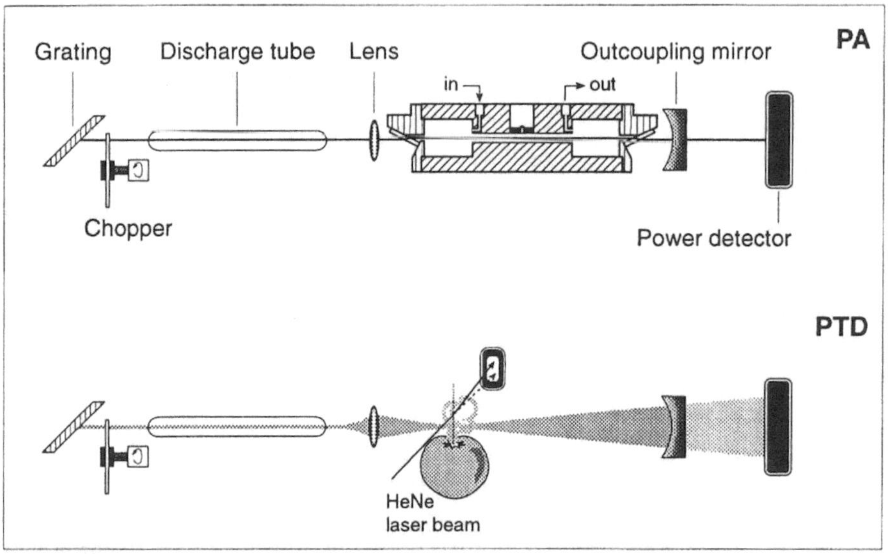

Fig. 3. *Above* The laser-driven photoacoustic system, consisting of a CO_2 laser and an intracavity-positioned photoacoustic (*PA*) cell. Ethylene concentrations are detected using three microphones placed on the resonator tube of the PA cell. *Below* Instead of a PA cell, a HeNe laser is used to measure ethylene concentrations near the plant tissue under normal atmospheric conditions. The deflection of this laser beam beam is monitored with a position-sensitive detector

pressure of 60 mbar. It consists of He, CO_2 and N_2, in a ratio of 65:13:22. At one end of the waveguide tube a grating is positioned as transition-selective element; a convex ZnSe lens (f = 75 mm) and a concave ZnSe outcoupling mirror (R = 270 mm) are placed at the opposite site. The distance between lens and mirror (i.e. the available space for investigated samples) is 350 mm. The cavity length, from grating to outcoupling mirror, is 1.2 m. The outcoupling mirror has a reflectivity of 98.4%, yielding the typical intracavity power of 100 W on the strongest lines, extracavity power is typically 5 W employing a 90% mirror. Due to the concave output mirror and the lens, the laser beam possesses a sharp intracavity focus with a waist of 0.28 mm. The grating, discharge tube, lens and outcoupling mirror are stably mounted in a single frame.

The HeNe laser passes at a distance of 0.7 mm over the heated region, i.e. above the focus of the CO_2 laser, employing several times a flat mirror multiple pass system; it consists of two square flat mirrors (flatness $\lambda/20$; Melles Griot) with highly reflective MAXBRIte coatings (R = 99.5%), tilted with respect to each other by a small angle Θ. In our actual experiments, the number of passes is 31 with $\Theta \approx 1°$ and a mirror distance of about 40 mm. The number of passes not only leads to an enhanced total deflection angle, but also increases the lever arm to the detector (de Vries et al. 1992); therefore, 31 passes lead to a 300-fold deflection as measured by the PSD.

For improved spatial resolution, two cylindrical mirrors are used instead of flat mirrors. Spherical mirrors would counteract the vertical deflection of the probe beam, whereas cylindrical mirrors do not influence the beam pass in that direction. The cylindrical mirrors tested had a focal length of 45 mm and are also coated with the MAXBRIte layer. The incoming HeNe laser beam is focused by a cylindrical lens (f = 76.2 mm; V-coating; Melles Griot) nearly at the center of the mirror setup with an offset of 1 mm. The experimental number of passes has been 15. The outcoming beam is made parallel by a second identical lens. Compared to the flat mirror setup, the probing length decreased from 25 to 2 mm length along the CO_2 laser; the vertical resolution of 0.56 mm (twice the waist of the CO_2 laser) remains unchanged. The sensitivity obtained with the cylindrical mirror setup as compared to the flat mirror setup decreases by a factor of three due to the actually restricted number of passes in the cylindrical mirror setup.

The probe beam, multipass system and detector are also mounted in a stable frame. The intracavity frame and the deflection frame are rigidly connected to minimize mechanical vibrations between the two laser beams. The entire setup is sustained by soft cushions to decouple external low frequency vibrations, especially important because a low chopper frequency of 20 Hz was used. Air turbulences are minimized by shielding the whole frame.

The entire experimental setup is computer controlled, in particular grating adjustment and piezo optimization of the cavity length, necessary to achieve single laser line operation. The positioning of the biological sample below the near crossing of the two laser beams is also adjusted by the computer. Deflection signal and intracavity CO_2 laser power, to which the signal is normalized, are stored for different laser lines. Concentrations of specific gases and the background signals are calculated. If desired, averaged values can be calculated including noise estimates. Measurements last typically 1 week.

2.2 Practical Considerations

The spatial resolution of the measurements is determined by the effective waist of the CO_2 laser and the probing length of the HeNe laser along the CO_2 laser. For our setup, this results in a spatial resolution of 0.56 mm diameter by 25 and 2 mm length in case of flat and cylindrical mirrors, respectively.

The radial dimensions of the CO_2 laser beam are of importance, considering possible temperature influences on plant tissues. In experiments, care has been taken to avoid e.g. heating and burning. In the wings of the Gaussian beam profile there is still a lot of laser power available. Remember that one has 100 W in a beam of 0.28 mm waist. In a simple experiment, we placed a thermocouple inside a tomato just beneath the epidermis, which showed that at least a distance of 1.5 mm has to be introduced between the center of the pump beam and the plant tissue, e.g. tomato surface. At this distance a temperature rise of 1 °C is then observed. In all our experiments we chose a distance of 1.5 mm. The temperature rise due to absorption of ambient gases (CO_2 and H_2O) is of the order of 10 to 100 mK at the center of the CO_2 laser beam. This temperature rise influences the tissue negligibly.

The PTD signal depends linearly on concentration, as was checked over the range of 12 to 1200 ppbv (parts per billion volume) of ethylene mixed in pure nitrogen. For this check, we started with a certified gas bottle of 1.2 ppmv (parts per million volume) C_2H_4 in pure nitrogen (supplied by UCAR) and applied further dilutions to a reference cell in the multipass mirror area.

In the case of ethylene, we deal with the strongly absorbing 10P14 line ($\alpha = 13.4\,atm^{-1}cm^{-1}$ at $949.479\,cm^{-1}$) and the 10P12 reference line ($\alpha = 4.31\,atm^{-1}cm^{-1}$ at $951.192\,cm^{-1}$). To account for partial saturation the quoted absorption coefficient on the 10P14 is taken lower than the value ($30.4\,atm^{-1}cm^{-1}$) given by Brewer et al. (1982). This, however, does not impair the linear dependence of the signal on the gas concentration. On the 10P14 laser line the PTD detection limit for ethylene is 0.5 nl/l. For ammonia this limit is 0.25 nl/l ($\alpha = 20.5\,atm^{-1}cm^{-1}$ at $967.708\,cm^{-1}$; Brewer and Bruce 1978).

Under ambient conditions gases such as CO_2 and H_2O are present, which also absorb CO_2 radiation (HITRAN data base 1992; Ryan et al. 1983). To eliminate their influence, we switch between two different CO_2 laser lines; CO_2 and the dimer of H_2O produce broad, unstructured absorptions. In our case, this is advantageous; variations of concentrations thus mainly result in a change in background signal: subtracting the signal for the reference laser line from the signal on the absorbing laser line leaves a small but still significant CO_2 contribution; e.g. 350 ppmv of ambient CO_2 corresponds to 4 ppbv of ethylene with this subtraction procedure. In plant experiments, besides CO_2 and H_2O, other gases can be present in high concentrations, e.g. ethanol, produced by fermentation in the absence of oxygen inside the plant tissue. Fortunately, also the absorption spectrum of ethanol is broad (Sigrist 1994), yielding an increase in background signal. Similarly, the role of broadband absorption of other carbohydrates like propylene leads mainly to an increase in background. In photoacoustic measurements, some of these components are cryotrapped, a trick that cannot be applied in open-air photothermal deflection measurements.

The time resolution is determined by switching between two different laser lines. Tuning the grating of the CO_2 laser has yielded a resolution of 15 s (1 s for single line operation). Employing a galvo-driven oscillating prism as line-selective element inside the CO_2 laser cavity, a resolution of 0.1 s has recently been obtained (de Vries et al. 1994a).

2.3 Local and Fast Ethylene Detection

2.3.1 Ethylene Emission Pathways for Intact Cherry Tomatoes

The PTD setup was first applied to investigate ethylene pathways of a single cherry tomato (*Lycopersicon esculentum* cv. Favorita). The fruit was in the mature orange stage, weight approximately 14 g and diameter about 35 mm. The tomato was detached from the plant and the calyx (often called coronet in the literature) was removed at least 10 h before the start of the experiment.

Measurements demonstrated that most of the internally produced ethylene (90%) is released at the stem-scar site (where the calyx had been situated) (Fig. 4). The amounts of ethylene emitted at the bottom and at the equator are both very low; measured levels are just above ambient concentrations of the order of 4 ppbv.

Our observation that 90% ethylene is emitted at the stem scar is in general agreement with the percentage obtained by preloading a normally sized tomato with high concentrations of ethane gas and determining the subsequent ethane efflux out of the stem scar and out of the rest of the fruit (Cameron and Yang 1982). Burg and Burg (1965) obtained a value of 60% for ethylene emission through the stem scar of a normally sized tomato by determining the total emission after sealing the stem scar with lanolin paste. However, blocking the stem scar is accompanied by oxygen deprivation of the fruit interior and, therefore, necessarily results in a perturbed metabolism (de Vries et al. 1994b).

2.3.2 Locally Wounded Tomato Tissue

Wounding is assumed to exert its effect at the step where SAM (=S-adenosylmethionine) is converted into ACC (1-aminocyclopropane-1-carboxylic acid), the direct precursor of ethylene. Normally this step, regulated by the enzyme ACC synthase, is rate-limiting in the cascade of events leading to ethylene production (Yang and Hoffman 1984). By means of "stress" the rate can be increased

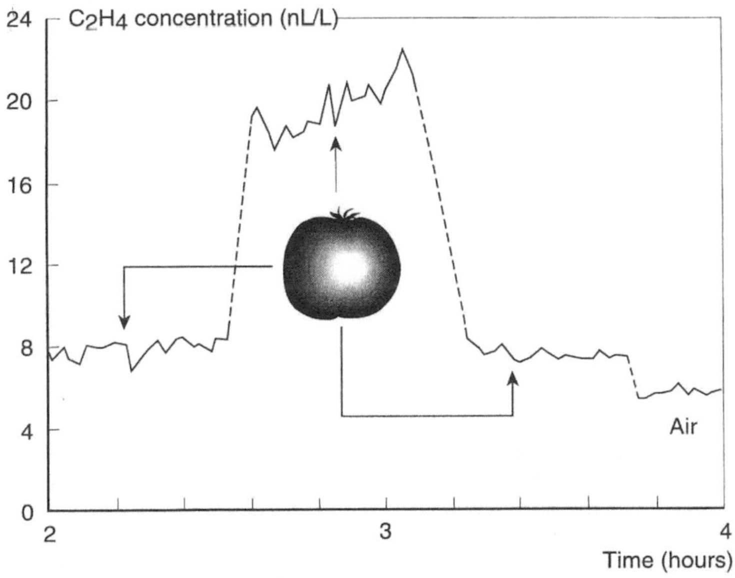

Fig. 4. Concentrations of ethylene were measured 1.5 mm above the surface of a green/orange tomato at various positions above an equatorial site, the stem scar and the bottom site, using PTD. For comparison, the background ethylene concentration in air without fruit is also shown (indicated by *Air*)

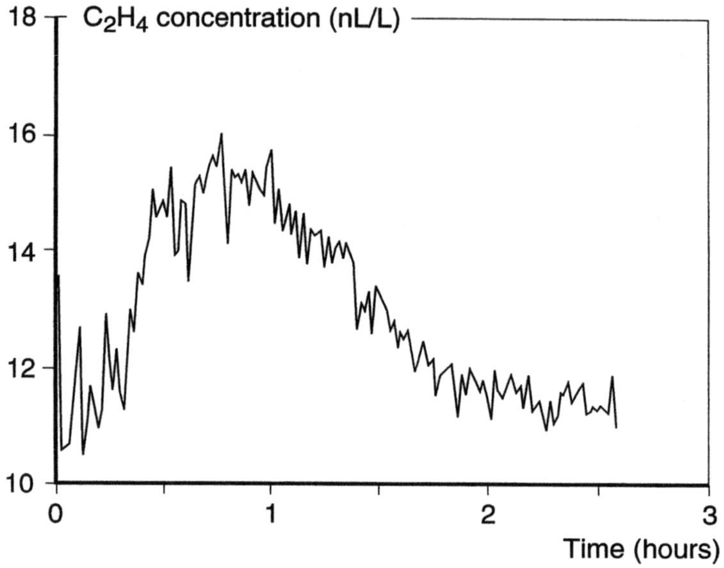

Fig. 5. Locally increased ethylene levels after removal of the calyx of a mature orange cherry tomato at t = 0, by PTD

(Boller and Kende 1980). The effect of calyx removal from a mature orange cherry tomato yields an increase in ethylene emission from t = 0 onwards lasting 2 h. Also, the local effect of wounding on the total ethylene emission at the stem-scar site and the bottom epidermis site was investigated in situ by PTD (Fig. 5; de Vries et al. 1994c). From these observations, it could be concluded that wounding the pericarp induces a much smaller effect than wounding the stem-scar site; this means that the wounding response differs locally for a single intact tomato fruit. It should be noted that the calyx-fruit junction is an abscission zone (Osborne 1989).

3 Photoacoustics Using a CO_2 Laser

3.1 The Experimental Setup

As radiation source for the PA measurements an infrared CO_2 waveguide laser was used in an intracavity configuration (100 W laser power) (see Fig. 3, above). A small resonant photoacoustic cell is placed inside the laser cavity between the output mirror and the waveguide tube. To reduce the laser beam radius so that no wall effects occur in the photoacoustic cell the laser beam is focused by a ZnSe lens (f = 254 mm) on a flat output mirror (reflectivity 92.1%). To generate the photoacoustic signal a mechanical chopper is placed inside the cavity. The photoacoustic cell possesses a resonator tube (length 100 mm, diameter 6 mm) which is resonantly excited at a frequency of 1600 Hz (Harren et al. 1990a). The thus produced acoustic signal is detected by three Knowles EK 3024 miniature microphones. On both sides of the resonator pipe buffer volumes are implemented to decrease the unwanted signal originating from (small) window absorption. Due to the small overall volume of the cell (34 ml), low flow regimes can be used and yield a relatively fast response time (1/e-time = 4 min at 1 l/h). The production levels of ethylene are often very low. High flow rates dilute concentrations. Therefore, low flows are used, although the time response declines and requires a compromise. A novel design yields an even better time response of 2 s at a flow of 5 l/h; in this case, the flow enters the resonator pipe centrally. Acoustic flow noise is reduced by quarter lambda dampers just before the entrance to the cell (Bijnen 1995).

A low flow regime is sometimes desirable because of small ethylene production rates (pl/h) by plant material. For the strongest vibrational absorption of ethylene on the 10P14 CO_2 laser line (α = 23.7 $atm^{-1}cm^{-1}$ at 949.479 cm^{-1}, see above), the detection limit is 6 pl/l (Harren et al. 1990a). To minimize background signals originating from interaction of the CO_2 laser beam with cell walls, reference measurements are performed on the neighbouring CO_2 10P12 laser line (α = 4.31 $atm^{-1}cm^{-1}$ at 951.192 cm^{-1}; Brewer et al. 1982); from these signals the ethylene concentration is calculated. In our laser photoacoustic setup we have a 1000 times higher sensitivity for ethylene than one conventionally copes with and, therefore, using a flowing air system we can avoid accumulation and pretreatment effects.

Ozone concentrations are measured with the same photoacoustic system using the 9P28 and 9P24 CO_2 laser lines (α = 8.7 $atm^{-1}cm^{-1}$ at 1039.369 cm^{-1} and α = 0.83 $atm^{-1}cm^{-1}$ at 1043.163 cm^{-1}, respectively; Harren et al. 1990b). A sensitivity of 20 pl/l is obtained. It has been checked that no saturation occurs for ozone.

3.2 Practical Considerations

The flow system coupled with the PA cell is shown schematically in Fig. 6. Before air enters the five cuvettes it is passed through a catalyst (platinized aluminium oxide pellets at 350 °C) to remove hydrocarbons as well as NO and NO_2. However, plant material in cuvettes 4 and 5 can also be fumigated by well-defined ozone concentrations in air or pure nitrogen; measurements can thus be performed under anaerobic and chemical stress conditions. The ozone is produced by a commercial electrical discharge ozonizer, fed with pure oxygen. The ozone-enriched oxygen is mixed with either purified air or nitrogen leading in two steps to the desired dilution. Starting with 3% ozone, oxygen-diluted concentrations of 10 to 10 000 nl/l are obtained. In a parallel fashion, through three additional cuvettes, purified air can be flowed. All cuvettes were continuously flushed; one of the flows through the cuvettes was selected to enter the PA cell. All flows are adjustable by flow controllers and are monitored by a mass flow sensor. For all experiments, ethylene levels of an empty cuvette were subtracted.

For the reproducibility of the measurements it is important that interfering gases such as CO_2 and H_2O are present constantly at low levels. Between the

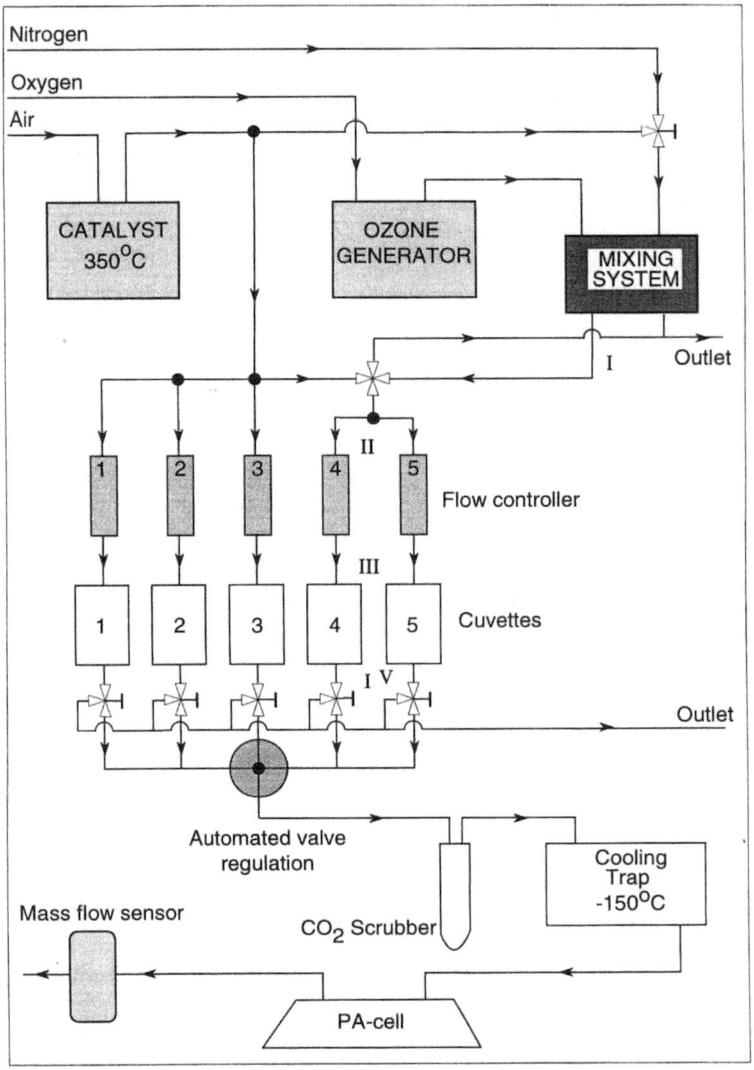

Fig. 6. The flow system coupled to the PA cell. Experiments can be performed using nitrogen or normal air - either with or without traces of ozone - passing a catalyst to remove hydrocarbons. All cuvettes are continuously flushed. Before gas enters the photoacoustic cell, scrubbers remove ozone, CO_2 and ethanol in case of ethylene measurements. In order to detect ozone these scrubbers are bypassed

sampling cell and the photoacoustic cell a KOH-based scrubber reduces CO_2 concentrations without influencing the ethylene concentration; a CO_2 concentration lower than 1 μl/l is reached, corresponding to an effective ethylene signal of 0.15 nl/l. For the experiments with pea seeds under anaerobic conditions, an additional cooling trap is placed between the sampling cell and the photoacoustic cell

(Petruzzelli et al. 1994). The cooling trap at −150 °C removes ethanol produced during the first hours of imbibition in high viability seeds (Woodstock and Taylorson 1981). The air passes through a wire netting at this temperature for optimal contact.

3.3 Ethylene Emission Rates for Various Species

3.3.1 Local Ethylene Emission by Bell Pepper

The measurements were performed on an intact bell pepper (*Capsicum annuum*) in the red stage. A distinction is made between emission rates through the calyx and through the skin. Thus, a cuvette consisting of two compartments is developed to sample air from calyx and skin separately. The calyx has been found to be the main emission avenue (70%) (Fig. 7). The results are comparable to those obtained with (cherry) tomatoes, both members of the same family (Solanaceae). A study of about ten different fruits has shown that all mainly emit ethylene through the skin except bell pepper and tomato (de Vries 1994).

3.3.2 Periodic Anoxic and Aerobic Conditions for Cherry Tomatoes

Withholding oxygen from a plant tissue is a direct measure to block ACC oxidase activity, thus stopping the ethylene production (Yang and Hoffman 1984; de Vries

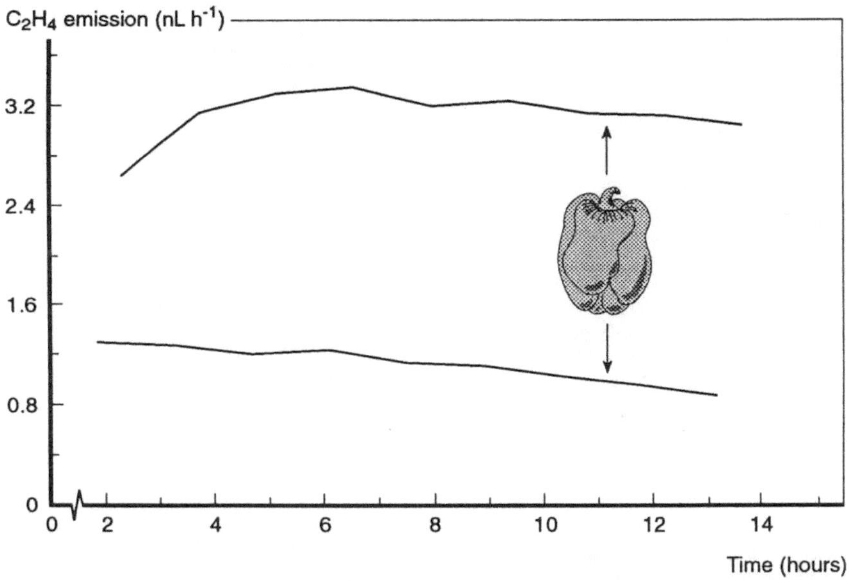

Fig. 7. Photoacoustically determined ethylene emission rates, employing the two-compartment cuvette, are presented for skin (*below*) and calyx (*above*) of a red bell pepper

et al. 1994a). In Fig. 8, the response of the ethylene emission from tomatoes is displayed for fast changes between anoxic and normal atmospheric conditions. In this case, gas is sampled only from the stem-scar region (main emission channel) to obtain a fast time response due to small sampling volumes. This figure demonstrates how fast the biosynthesis of ethylene is resumed after lack of oxygen and how fast it is inhibited. It also shows how well the PA technique is suited to follow rapid physiological changes employing a novel PA cell with a time response of 2 s (flow 5 l/h; Bijnen 1995). It should be noted that the PTD technique is at least one order of magnitude faster, as has been shown for ammonia measurements in the atmosphere (de Vries et al. 1995).

3.3.3 Germinating Peas Periodically Fumigated with Ozone

Ten peas (*Pisum sativum* L.) were fumigated with a 90 nl/l ozone-air mixture every 24 h for a period of 7 h (Fig. 9). During the first of three fumigation periods, 6 h after the start of the experiment, the production level increased rapidly and stayed high for the time of the fumigation period. Back to normal air the ethylene production dropped to zero. The start of a second fumigation period of 8 h at t = 23 h was accompanied by visible root formation and showed little extra effect on the ethylene production. The third fumigation period showed again a clear rise in production. The non-fumigated period in between had a low production level. The integrated ethylene production rate was lower than for the control peas and the

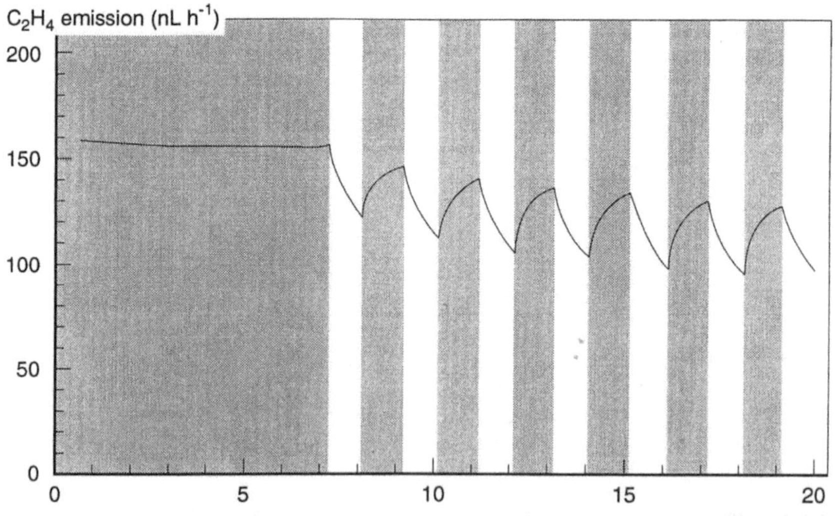

Fig. 8. Fast changes between anoxic and aerobic conditions show fast changes in ethylene emission rates (PA). The measurements start with an aerobic period of 7.2 min whereafter an orange-red cherry tomato is exposed alternatingly to nitrogen or air, for 1 min. In addition to the very fast ethylene response, a slow decay is observed yielding an average ethylene emission

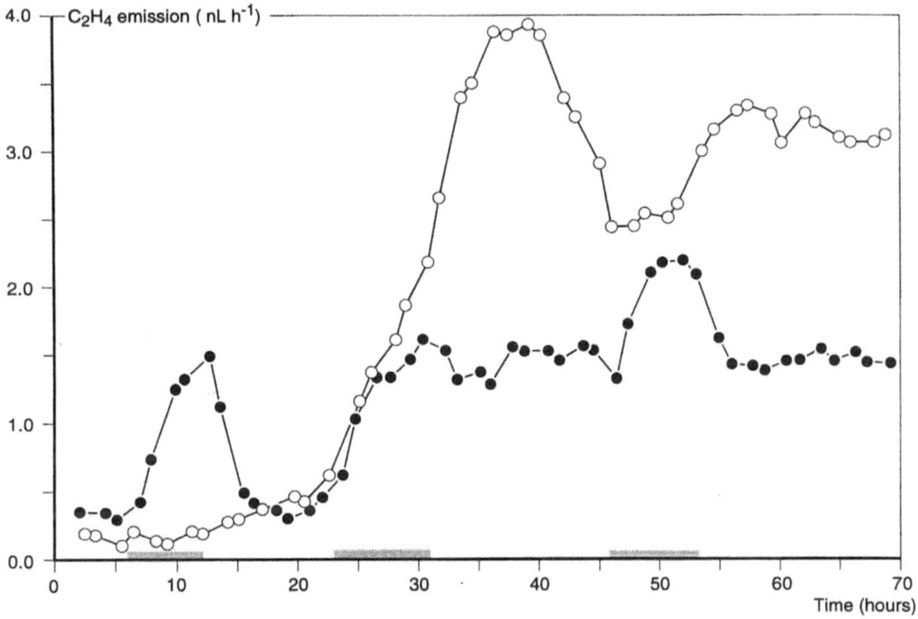

Fig. 9. Ten peas periodically fumigated with 90 nl/l ozone (•); the fumigation periods are indicated; levels of non-fumigated peas are also shown (○)

germination percentage was 50%; in general, for control and periodically fumigated peas the germination percentages were similar. In this contribution, the conclusion is drawn that ethylene production is enhanced by short ozone fumigation periods, even before visible germination takes place. Such an increase is also observed for *Striga* seeds treated either by strigol analogue GR24 (or by ACC; Thuring et al. 1994). It should be noted that one of the first examples of pea seedlings fumigated with ozone is presented by Mehlhorn and Wellburn (1987).

3.3.4 *Rumex palustris* Exposed to Flooding Stress

The last example deals with submerged *Rumex* species. A hypoxic atmosphere thus exists in the tissue because gas diffusion is limited in water. In contrast to *Rumex acetosa*, *Rumex palustris* shows stimulated shoot elongation in response to submergence paralleled and triggered by drastic enhancement of ethylene concentration. This elongation response is functionally related to the opportunity to reach better illuminated and aerated zones close to the water surface or preferentially above it (Voesenek and van der Veen 1994). Desubmergence of the species displays a first peak in ethylene emission with a subsequent second rise followed by a slow decay (Fig. 10). As discussed by Voesenek et al. (1993), the first peak is due to release of entrapped ethylene. The second rise corresponds to a temporary

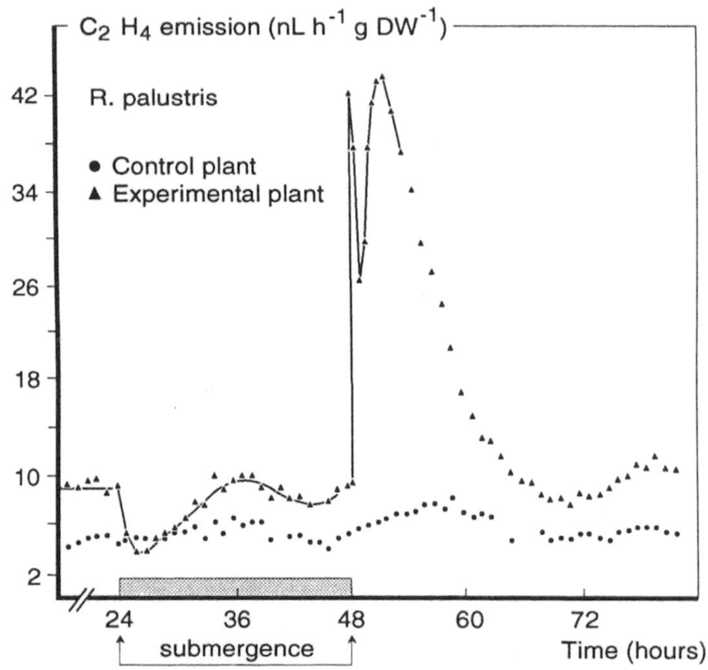

Fig. 10. Ethylene emission rates from shoots of *Rumex palustris* before, during and after a submergence period indicated by the *shaded bar*. Control plants under non-submerged conditions show a constant rate

increase in ethylene synthesis. Remarkably, a similar pattern is observed for tomatoes and germinating peas returning from anaerobic conditions; however, in those cases, the first peak represents conversion of accumulated ACC (de Vries et al. 1994a). It should be noted that *Rumex palustris* survives high water levels in lower parts of the river areas, whereas the habitat of *Rumex acetosa* is situated in higher domains seldomly reached by rising water.

4 Photoacoustics Using a CO Laser

4.1 The Experimental Setup

Instead of using a CO_2 laser as excitation source in the PA scheme, a CO laser offers the opportunity to measure trace gases (Bijnen et al. 1993). The CO laser emits radiation between 1260 and 2080 cm^{-1} at about 250 transitions ($\Delta v = 1$). Single line lasing is achieved by cooling the discharge tube of the laser with liquid nitrogen. To increase the available laser power to 40 W the PA cell is positioned in the cavity of the laser. Care is taken to avoid interaction of laser radiation and cell walls by implementing a 100% reflecting silver-coated hollow end mirror (R = 5 m). The laser beam is focused on the grating placed at the opposite side of the cavity. The grating can be automatically tuned to achieve lasing at different laser lines. Various gases possessing specific fingerprint absorption spectra on these lines can thus be analyzed; background signals can be subtracted. The time response of this setup is 15 s at a flow of 5 l/h for single laser line operation; the design of the PA cell is similar to that discussed in Section 3.

4.2 Practical Considerations

The PA cell is coupled with a flow-through system. Special attention is paid to the reduction of water vapour concentrations in the PA cell; high levels interfere with trace gas detection because of high absorptions covering a large part of the CO laser spectrum. Thus, either a cryotrap at T = −150 °C or small columns filled with desiccated P_2O_5 are used between sampling cuvette and PA cell. At well-defined temperatures cryotraps were also installed to gain spectral selectivity for other trace constituents.

With this equipment many small molecules are detectable (Table 1). Some of them are particularly important in fruit analysis. Water vapour detection allows one to monitor respiration processes; ethanol and especially acetaldehyde measurements permit one to follow the onset of fermentation (Bijnen 1995); ethane is taken as an indicator of cell membrane damage. It should be noted that periodic release of methane by cockroaches is investigated in detail with this setup (Bijnen 1995).

5 Résumé

PTD and PAD techniques are developed for non-intrusive detection and quantification of trace gases. Detection limits and time responses of 6 pl/l and 2 s for PAD

Table 1. For the CO laser, extrapolated detection limits in pure N_2 are given for biologically interesting gases, based on absorption coefficients and laser power. (Bijnen et al. 1994)

Trace gas	Detection limit	Trace gas	Detection limit
CS_2	10 pl/l	CH_4	1 nl/l
CH_3COOH	100 pl/l	$S(CH_3)_2$	1 nl/l
H_2O	100 pl/l	$N(CH_3)_3$	1 nl/l
NO_2	100 pl/l	CH_3CHOH	3 nl/l
SO_2	100 pl/l	$CH_3(CH_2)_3CH_3$	3 nl/l
N_2O	1 nl/l	CH_3SH	10 nl/l
C_2H_2	1 nl/l	H_2S	1 µl/l
C_2H_6	1 nl/l	CO_2	1 µl/l
C_2H_4	1 nl/l		

Table 2. PTD and PA applications in physiological processes that involve ethylene production

Natural processes	Species	Ethylene production (E.P.)	Reference
Germination	Peas	E.P. accompanies germination	Petruzzelli et al. (1994)
		E.P. and germination retarded by decoating	
		E.P. inhibited by AVG while normal germination is observed	de Vries (1994)
	Striga	E.P. and germination induced by CR24 or ACC	Thuring et al. (1994)
Growth	*Rumex*	E.P. and growth enhanced by flooding (species-specific); E.P. doubly peaked after desubmergence	Voesenek et al. (1993) Harren et al. (1990a)
	Potamogeton	E.P. not observed	(J.E. Summers, M.B. Jackson and L.A.C.J. Voesenek, pers. comm. 1994)
Wilting	Orchids	E.P. doubly peaked after emasculation (drought stress)	Woltering et al. (1988)
Ripening	Tomatoes	E.P. shows climacteric pattern	Bijnen et al. (1994)
	Apples/tomatoes bananas/kiwi etc.	E.P. per surface area changes highly, species-dependent for different sites, during ripening	de Vries (1994)

Stress effects	Species	Ethylene production (E.P.)	References
Anoxia	Peas/tomatoes	E.P. doubly peaked after restored aerobic conditions	de Vries et al. (1994a,b)
Wounding	Tomatoes	E.P. doubly peaked, differs locally for calyx and skin	de Vries et al. (1994c)
	Bell pepper	E.P. single peak, differs locally for calyx and skin	de Vries et al. (1994d)
Chilling	Peas	E.P. and germination retarded by 8 h	Petruzzelli et al. (1994)
	Tomatoes	E.P. retarded by 8 h	de Vries (1994)
Ozone	Peas	E.P. even before visible germination, periodic effects	de Vries (1994)
Flooding	*Rumex*	See above	
Drought	Orchids	See above	

and of 0.5 nl/l and 0.1 s for PTD are achieved. The CO_2 laser-driven photothermal deflection instrument based on the mirage effect makes it possible to detect local concentrations, e.g. of ethylene and ammonia, down to 0.5 and 0.25 ppbv, respectively, in the atmosphere. The time response is 0.1 s. A detection limit of 6 pptv (parts per trillion volume) for ethylene and a temporal resolution of 1 s are reached employing a photoacoustic setup coupled with a flow-through system. Using a CO laser instead of a CO_2 laser in the photoacoustic configuration, gases such as methane, ethane, acetaldehyde, water and dimethylsulfide are detected at nl/l levels.

The features of the new sensitive, local and fast laser-based detection scheme are elucidated by a study of several biological processes that involves ethylene production. A complete list of PTD and PA applications is given in Table 2.

Acknowledgements. This contribution is a review of applied laser-photoacoustic and -photothermal research performed under the supervision of Dr. F.J.M. Harren and Prof. Dr. J. Reuss, in collaboration with Dr. F.G.C. Bijnen, all from the group of Molecule and Laser Physics. Dr. L.A.C.J. Voesenek and Prof. Dr. C.W.P.M. Blom (Department of Ecology, University of Nijmegen), Dr. E.J. Woltering and Dr. H.C.P.M. van der Valk (ATO-DLO, Wageningen) and Dr. L. Petruzzelli (Instituto Germoplasma, Bari) contributed at various stages of this work. More details can be found in the PhD Thesis of the author (Nijmegen 1994).

References

Abeles FB, Morgan PW, Saltveit ME Jr (1992) Ethylene in plant biology, 2nd end. Academic Press, London

AFGL-HITRAN data base, USF HITRAN-PC Version (1992) (ed Killinger K) Univ of South Florida, Tampa

Bassi PK, Spencer MS (1989) Methods for the quantification of ethylene produced by plants. In: Linskens HF, Jackson JF (eds) Modern methods of plant analysis, vol 9. Springer, Berlin Heidelberg New York, pp 309–321

Bijnen FGC (1995) Refined CO-laser photoacoustic trace gas detection, applied to insects. PhD Thesis, University of Nijmegen, The Netherlands

Bijnen FGC, Harren FJM, Reuss J, van Hoek AHAM, van Alen TA, Hackstein JHP (1993) Intracavity CO-laser photoacoustic trace detection; CH_4 production by methanogenic bacteria. In: Werner Ch, Weiderlich (eds) Lasers in remote sensing. Springer, Berlin Heidelberg New York, pp 66–71

Bijnen FGC, de Vries HSM, Harren FJM, Reuss J (1994) Cockroaches and tomatoes investigated by laser photoacoustics. J Phys IV, Colloq C7 (Suppl J Phys III) 4:435–443

Boller T, Kende H (1980) Regulation of wound ethylene synthesis in plants. Nature 286:259–260

Brewer RJ, Bruce CW (1978) Photoacoustic spectroscopy of NH_3 at the 9-μm and 10-μm $^{12}C^{16}O_2$ laser wavelengths. Appl Opt 21:4092–4100

Brewer RJ, Bruce CW, Mater JL (1982) Opto-acoustic spectroscopy of C_2H_4 at the 9 and 10 μm CO_2 laser wavelengths. Appl Opt 21:4092–4100

Burg SP, Burg EA (1965) Gas exchange in fruits. Physiol Plant 18:870–884

Cameron AC, Yang SF (1982) A simple method for the determination of resistance to gas diffusion in plant organs. Plant Physiol 70:21–23

de Greef JA, de Proft M (1978) Kinetic measurements of small ethylene changes in an open system designed for plant physiological studies. Physiol Plant 42:79–84

de Vries HSM (1994) Local trace gas measurements by laser photothermal detection; Physics meets physiology. PhD Thesis KUN, Nijmegen, The Netherlands de Vries HSM, Harren FJM, Reuss J (1992) The photothermal deflection technique (PTD): fast trace gas detection in the atmosphere. In: Bicanic D (ed) Photoacoustic and photothermal phenomena III. Springer, Berlin Heidelberg New York, pp 12–15

de Vries HSM, Bijnen FGC, Voesenek LACJ, Blom CWPM, Harren FJM, Reuss J (1994a) Laser photothermal deflection applied to local trace gas detection; respiration of tomatoes. Laser Optoelektronik (in press)

de Vries HSM, Harren FJM, Voesenek LACJ, Blom CWPM, Woltering EJ, van der Valk HCPM, Reuss J (1994b) In situ investigation of ethylene emission for intact cherry tomatoes by means of photothermal deflection and photoacoustic detection. Plant Physiol (in press)

de Vries HSM, Harren FJM, Reuss J (1994c) In situ, real-time monitoring of wound-ethylene in cherry tomatoes by two infrared laser-driven systems. POSTHARVEST Biology and Technology (in press)

de Vries HSM, Harren FJM, Woltering EJ, van der Valk HCPM, Reuss J (1994d) Laser investigation of ethylene emission by fruits. Proc Cost 94 Conf, Oosterbeek, The Netherlands, Nov 1994

de Vries HSM, van Lieshout MR, Harren FJM, Reuss J(1995) Infrared laser photothermal trace gas detection applied to environmental and biological problems. Infrared Phys 36:483–488

Harren FJM, Bijnen FGC, Reuss J, Voesenek LACJ, Blom CWPM (1990a) Sensitive intracavity photoacoustic measurements with a CO_2 waveguide laser. Appl Phys B50:137–144

Harren FJM, Reuss J, Woltering EJ, Bicanic DD (1990b) Photoacoustic measurements of agriculturally interesting gases and detection of C_2H_4 below the ppb level. Appl Spectrosc 44:1360-1367

Huelin FE, Kennett BH (1959) Nature of the olefins produced by apples. Nature 184:996

Jackson WB, Amer NM, Boccara AC, Fournier D (1981) Photothermal deflection spectroscopy and detection. Appl Opt 20:1333–1344

Kende H (1993) Ethylene biosynthesis. Annu Rev Plant Physiol Plant Mol Biol 44:283–307

Mehlhorn H, Wellburn AR (1987) Stress ethylene formation determines plant sensitivity to ozone. Nature 327:417–418

Osborne DJ (1989) Abscission. Crit Rev Plant Sci 8:103–129

Petruzzelli L, de Vries HSM, Harren FJM (1994) Alleviation of injury due to a short cold exposure in germinating peas by seed decoating and its effects on ethylene synthesis and respiration. (Submitted)

Ryan JS, Hubert MH, Crane RA (1983) Water vapor absorption at isotopic CO_2 laser wavelengths. Appl Opt 22:711–717

Seymour GB, Taylor JE, Tucker GA (1993) Biochemistry of fruit ripening. Chapman and Hall, London

Sigrist MW (1994) Air monitoring by spectroscopic techniques. In: Winefordner JD (ed) Chemical analysis 127. John Wiley, New York

Thuring JWJF, Harren FJM, Nefkens GHL, Reuss J, Titulaer GTM, de Vries HSM, Zwanenburg B (1994) Ethylene production by seeds of *Striga hermonthica* induced by germination stimulants. Proc 3rd Int Worksh on Orobanche (8–12 Nov), Amsterdam

Voesenek LACJ, van der Veen R (1994) The role of phytohormones in plant stress: too much or too little water. Acta Bot Neerl 43:91–127

Voesenek LACJ, Banga M, Thier RH, Mudde CM, Harren FJM, Barendse GWM, Blom CWPM (1993) Submergence-induced ethylene synthesis, entrapment, and growth in two plant species with contrasting flooding resistances. Plant Physiol 103:783–791

Woltering EJ, Harren FJM, Boerrigter HAM (1988) Use of a laser-driven photoacoustic detection system for measurement of ethylene production in *Cymbidium* flowers. Plant Physiol 88:506–510

Woodstock LW, Taylorson RB (1981) Ethanol and acetaldehyde in imbibing soybean seeds in relation to deterioration. Plant Physiol 67:424–428

Yang SF, Hoffman NE (1984) Ethylene biosynthesis and its regulation in higher plants. Annu Rev Plant Physiol 35:155–189

Induction of Fruit Aroma and Quality by Post-Harvest Application of Natural Metabolites or Anaerobic Conditions

E. Pesis

1 Biosynthesis and Degradation of Aroma Volatiles During Post-Harvest Life

1.1 Ripening and Aroma Production in Fruits

During fruit ripening, major changes occur, some of which are degradative, e.g. chlorophyll breakdown, starch hydrolysis and cell-wall degradation, and some constructive, such as formation of carotenoids and anthocyanins, synthesis of aroma volatiles, and ethylene formation (Biale and Young 1981). The increase in aroma volatiles contributes to the flavour, taste and smell of the fruits. Each kind of fruit has its unique assemblage of volatile compounds, divided into main groups of acids, alcohols, esters, carbonyls and hydrocarbons (Nursten 1970). Acetyl coenzyme A (CoA) is the precursor for the acetate esters, while the acyl-CoA is the precursor for the longer esters and acts as the acylating agent in ester formation (Gilliver and Nursten 1976).

In both climacteric and non-climacteric fruit the most important aroma volatile compounds that increase during ripening are the esters. In Bartlett pears, during shelf life, the climacteric rise in CO_2 and ethylene was accompanied by the production of huge amounts of volatiles; acetaldehyde (AA), methyl acetate, ethyl acetate and propyl acetate increased throughout the ripening and softening process, but butyl acetate and amyl acetate increased with the climacteric rise and dropped with climacteric decline (Romani and Ku-Lim 1966). In banana fruit there was a dramatic increase in various esters, mainly acetate esters, including *n*-propyl acetate, ethyl acetate, *iso*-butyl acetate, *n*-butyl acetate and *iso*-amyl acetate, and butyrate esters including ethyl butyrate, *n*-butyl butyrate, *iso*-butyl butyrate and *iso*-amyl butyrate (Tressl and Jennings 1972). During ripening of banana at 25 °C, the increase in acetate and butyrate esters was correlated with increases in ethanol and 3-methyl butanol, respectively, and after 6–8 days a gradual decline in ester production began (Tressl and Jennings 1972). A linear correlation between the ratio of acetate esters to butyrate esters and ripening time was found in bananas ripened at 20 °C (Macku and Jennings 1987).

Feijoa are very aromatic fruit; they have their best aroma and flavour after natural abscission, but during storage they lose their special flavour. Shaw et al. (1983) reported a decrease in several volatiles, mainly ethyl butyrate, and an

Modern Methods of Plant Analysis, Vol. 18
Fruit Analysis
Edited by H.-F. Linskens and J.F. Jackson

increase in methyl benzoate and ethyl benzoate production in intact feijoa fruit, during 15 days of storage at 17 °C. The processes of synthesis and breakdown are probably responsible for the variations in aroma and flavour.

During tomato fruit ripening, at 21 °C, from the green stage to the red stage, only the flavour volatile, eugenol, decreased; ethanol and *trans*-2-*trans*-4-decadienal levels showed no change as the fruit ripened, while most of the aroma volatiles (acetaldehyde, *cis*-3-hexenal, *trans*-2-hexanal, hexanal acetone, 6-methyl-5-hepten-2-one, geranylacetone and 2-isobutylthiazole) increased in concentration, peaking in the turning, pink and red stages of maturity. The volatile, *cis*-3-hexanol, increased during ripening and reached its peak in the pink stage but subsequently, in the red stage, it suffered a dramatic decrease (Baldwin et al. 1991).

In mango fruit, the characteristic flavour appeared to develop only after the half-ripe stage (the climacteric stage) and the extent of flavour generation depended on the temperature conditions during storage (Gholap et al. 1986).

Also in non-climacteric fruits, such as strawberry and pineapple, there is a dramatic increase in aroma volatiles. The contribution of esters as well as acetals, alcohols and aldehydes to the strawberry aroma has been pointed out by Teranishi et al. (1963). In pineapple there is a dramatic increase in ester production during ripening (Gortner et al. 1967). On the other hand, in citrus there are no dramatic changes in volatile production during storage. In grapefruit and pomelo, the changes in volatiles are very minor during storage, except for the important increase in nootkatone which contributes significantly to pleasant grapefruit flavour (Pino et al. 1986; Sawamura et al. 1989).

1.2 Measurements of Sensory Quality

Food flavour is perceived as a result of the release into the air of volatile odour components, usually present in food only in trace amounts. Direct sampling of volatiles emanating from fruits by means of gas chromatography (GC) is one of the best non-destructive means to assess fruit ripeness. In order to correlate the contribution of certain volatile constituents to the flavour, it is possible to use the odour threshold value and to compare it with the GC measurements. Ahmed et al. (1978) demonstrated that the odour and flavour thresholds are identical. Odour threshold concentrations can be determined by using the single-stimulus-difference test employed by Siek et al. (1969). The test involved presenting the members of a panel with several samples, in different dilutions, together with a water standard for reference. The statistical analyses for determining the threshold values involved prediction of the concentration that corresponded to 50% positive responses among all the panel reactions judgements (Ahmed et al. 1978). The relative flavour contribution of each volatile is calculated as the ratio of the amount present in the fruit juice and the odour threshold value of each compound (Pino et al. 1986). Comparison of the relationships between sensory responses and GC data, in conjunction with odour threshold concentrations, revealed which volatiles contribute to pleasantness or unpleasantness of grapefruit flavour. In grapefruit juice, the highest relative flavour contribution among 34 volatile compounds was that of ethyl butyrate, followed by that of ethyl acetate (Pino et al. 1986).

2 Induction of Aroma Volatile Production by Application of Aldehydes, Alcohols and Acids

2.1 Application of Acetaldehyde

Acetaldehyde (AA), a natural aroma component in almost every fruit, accumulates during ripening (Fidler 1968). Acetaldehyde in both climacteric (Dilley 1970) and non-climacteric (Yamashita et al. 1978) fruits is formed from pyruvate by the enzyme, pyruvate decarboxylase (PDC). The two immediate products formed from AA are: (1) ethanol, by the enzyme, alcohol dehydrogenase (ADH), and (2) acetyl CoA by the enzyme, aldehyde dehydrogenase (Cossins 1978).

Application of AA vapour to whole fruits enclosed in jars was performed by flushing air over liquid AA, to produce air enriched with AA vapour. Different AA concentrations were achieved by restricting the flow of the AA source through a series of glass capillaries. The AA vapour was mixed with air enriched with water vapour in order to maintain a suitable humidity in the fruit jars and to prevent drying of the commodity (Fig. 1).

During the treatments, head-space gas samples were taken from the jars, and AA concentration was measured against known gas standards by GC, using a 10% Carbowax 20 M column, flame ionization detector at 180 °C, N_2 carrier gas, injection temperature of 110 °C, and oven temperature of 80 °C. After the treatment, the jars were ventilated for up to 24 h with humidified air and then opened, while control fruits were ventilated with humidified air alone for 24 h. The fruits were transferred to cardboard boxes for further cold storage or for shelf life (Pesis and Avissar 1989).

Other aldehydes, alcohols and acids were applied as vapours in similar ways. For example, apples (1.5–2 kg) were placed in desiccators (8 l). The vessels were continuously flushed with air. Test substances (100 μl) which were sufficiently volatile at room temperature to ensure complete evaporation in 24 h (acetic, propionic and butyric acids, propanal and butanal) were placed in a two-necked flask, installed between a compressed air cylinder and the desiccator, and carried into the apple containers as they evaporated (De Pooter et al. 1983). The less volatile compounds (pentanoic and hexanoic acids; pentanal and hexanal) were injected directly into the incubation vessel (De Pooter et al. 1983).

Application of the compounds can also be done in a closed system. For example, 2 kg tomato fruits were placed in 20-l jars, the fruits were exposed to alcohol

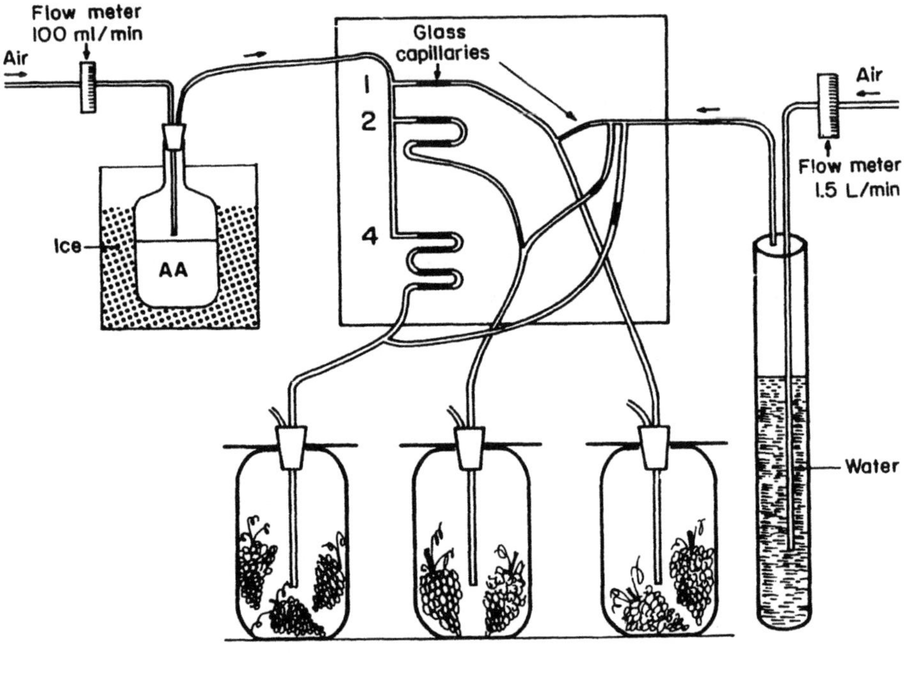

Fig. 1. Schematic of acetaldehyde (AA) generator which supplies various concentrations of AA via glass capillaries. At the end of the treatment the flow over the AA is blocked and the fruit in the jars continues to receive humidified air until no AA residue remains in the head-space atmosphere

vapours by applying alcohol to a 6 × 35-cm strip of filter paper, taped at one end to the rim of the jar. The other end of the paper strip rested on the bottom of a beaker in the jar so that the liquid alcohols did not come into contact with the fruit (Kelly and Saltveit 1988).

Fruit discs, fruit slices or small fruits held in small glass flasks sealed with a rubber cap can serve as a simplified model for studying volatile formation by applying various compounds as gas or liquid (Yamashita et al. 1975; Gilliver and Nursten 1976; Berger and Drawert 1984).

2.2 Biosynthesis of Aroma Volatiles from Various Precursors

Application of AA vapour has been shown to lead to major changes in fruit-ripening processes in non-climacteric and climacteric fruit (Fidler 1968). Yamashita et al. (1975, 1976) showed that intact strawberries were able to synthesize carboxylic esters on addition of alcohols and acids or aldehydes and acids. In addition, aldehydes were reduced to the corresponding alcohols, e.g. ethanol and ethyl esters were formed from AA; 1-propanol and 1-propyl esters were formed from propanal; 1-butanol and butyl esters were formed from 1-butanal (Yamashita et al. 1976).

Extensive work has been done to increase apple aroma artificially, by increasing the amounts of volatile esters in the fruit. Golden Delicious apple fruits treated with acetic, propionic and butyric acids showed an increase in the corresponding esters (De Pooter et al. 1981, 1983). There was a gradual increase in ester formation in both the control and treated fruits until 15 days after treatment. Later, in storage, there was a reduction in the level of esters (De Pooter et al. 1983). Berger and Drawert (1984) showed that storage of intact Red Delicious apples in an atmosphere containing ethanol vapours for 24 h resulted in a more than threefold increase in the ethyl ester concentrations. Bartley et al. (1985) showed that alcohols applied to apples were converted to the corresponding esters. The biosynthesis of ethyl acetate and ethyl butyrate in fruits is probably preceded by ethanolysis of acetyl-CoA and butyryl-ACP, respectively (Nursten 1970).

De Pooter et al. (1987) demonstrated the ability of Golden Delicious apples to produce aldehydes from exogenously supplied carboxylic acids, which led to the formation of volatile esters. Cold-stored apples efficiently transformed the aldehydes into volatile esters, but controlled-atmosphere (CA) stored fruits seemed to be unable to handle increased concentrations of aldehydes, probably because of changes in the ADH activity in the fruit (De Pooter et al. 1987).

2.3 Flavour and Taste Enhancement

Flavour enhancement after AA or ethanol application has been demonstrated in pears, tomatoes and blueberries (Paz et al. 1982), grapes (Pesis and Frenkel 1989), strawberry (Pesis and Avissar 1990) and feijoa (Pesis et al. 1991). In strawberry, a gradual increase in AA, ethanol, methyl acetate, ethyl acetate and ethyl butyrate production after application of AA was correlated with the duration of application (Table 1). The longer the application time, the greater the production of volatiles. However, excessive volatile production in the longer applications caused off-flavours (Pesis and Avissar 1990). The taste of the strawberries treated with AA for only 1 h was better than that of other treated or untreated strawberries, and the largest group of judges in the test panel (50%) preferred these strawberries (Table 1).

On the other hand, in cirtrus fruits it is difficult to improve fruit taste by exogenous treatment. Application of AA to Hamlin oranges caused induction of ethyl butyrate, but there was no major difference in the taste of the fruit juice (Shaw et al. 1991). Also in gamma-irradiated grapefruit there were significant changes in sugars and acids but there were no changes in the juice flavour (Nunez Selles et al. 1986).

Table 1. Effect of application of acetaldehyde (AA) vapour (5000 ppm) for 1 to 4 h on taste preference, and the accumulation of AA, ethanol, ethyl acetate, methyl acetate and ethyl butyrate in strawberry juice, 1 day after treatment at 20 °C

AA treatment (h)	Volatiles accumulation (ppm)					Preference (%)
	AA	Methyl acetate	Ethyl acetate	Ethanol	Ethyl butyrate	
0	4.2	17.3	20	30	2.5	20
1	15.6	29.4	280	52	16.2	50
2	19.4	32.5	520	105	17.5	10
3	26.5	32.8	850	411	22.6	5
4	31.0	31.3	910	546	21.4	15

2.4 Effect of Acetaldehyde on Sugar and Acid Content

Post-harvest application of AA to non-climacteric fruits has been reported to cause induction of CO_2 production (a climacteric-like respiration) in orange (Fidler 1968; Pesis and Avissar 1989), fig (Hirai et al. 1968), strawberry and blueberry (Janes et al. 1978) and grape (Pesis and Marinansky 1992). In fig and orange, AA application leads to reduced acidity (Hirai et al. 1968; Pesis and Avissar 1989).

In fig, AA caused acceleration of ripening, leading to more accumulation of sugars (Hirai et al. 1968). Application of AA vapour to blueberries, tomatoes and pears led to the enhancement of the fruit sensory quality, including an increase in the sugar content, sugar/acid ratio, and flavour changes which were judged to be acceptable by test panelists (Paz et al. 1982). In Sultanina and Perlette grapes with initially low sugar content (13–14% TSS) and high acidity, AA treatment increased TSS, decreased acidity, and enhanced the sensory preference by judges in a test panel (Pesis and Frenkel 1989). In citrus fruits, AA vapour caused yellowing of the peel and reduced fruit acidity, probably because of higher respiration rates during treatment.

In tomato fruits, AA led to a significant increase in the reducing sugars and sucrose, when applied at the red ripeness stage (Poole 1983). The possible stimulation of gluconeogenesis by AA vapour was demonstrated with radioactive work on tomato discs. The lable from applied L-[U-^{14}C] malic acid was recovered in respiratory CO_2, in other organic acids, in ethanol-insoluble material and, in an appreciable quantity, in the ethanol-soluble sugar fraction (Halinska and Frenkel 1991). The carbon flux from the acid to sugars in tomato pericarp tissue and the AA-stimulation of the process was accompanied by attenuation of the fructose 2,6-biphosphate levels (Halinska and Frenkel 1991).

2.5 Volatile Formation in Vitro

Although it is very easy to induce production of ester volatiles by applying various aldehydes, acids or alcohols to whole fruit or discs, it is very difficult to produce these volatiles in the test tube without the presence of the intact plant tissue.

When whole strawberries were cut into pieces and incubated with alcohols and acids, it has been shown that if they were cut into two or four pieces, the formation of the acetate esters increased, while cutting them into eight pieces dramatically depressed the formation of esters. Moreover, no ester formation was observed when strawberry was homogenized (Yamashita et al. 1975). Gilliver and Nursten (1976) showed that without adding banana slices, it was impossible to produce esters in aqueous solutions containing lower alcohols and either acetyl or butyryl CoA, or acetic or butyric anydride. They concluded that there is need for enzymes and/or cofactors from the fruit slice for ester biosynthesis.

However, in the last decade several papers were published on volatile and ester production in vitro. In mango cv. Alphonso, the fruity flavour could be regenerated in vitro by using acetone powder enzyme mixtures, substrate from the acetone filtrate and acetaldehyde (Gholap et al. 1986). Alcohol acyl-CoA transferase (ester synthetase) was extracted from the pulp of banana, melon and strawberry. A high reactivity of the enzyme from the three fruits was observed when propyl, isobutyl, butyl and isoamyl alcohols were used and reacted well with three acyl-CoAs: acetyl CoA, propionyl CoA and butyryl CoA (Ueda et al. 1992). In apples, after removal from 9 months of refrigerated air storage or CA storage, acetyl CoA alcohol transferase (ACAT) activity was demonstrated by esterification of butanol to butyl acetate in the presence of acetyl CoA and butanol (Fellman et al. 1993). At the day of removal to shelf life ACAT activity was higher in the cold-stored fruit than in the CA-stored fruit. However, the CA-stored apple showed a substantial increase in ACAT activity during shelf life. The ACAT protein could be either reactivated by exposure to air or synthesized de novo (Fellman et al. 1993).

There is a demand in the food industry for natural flavours which are produced by biotransformation of natural sources into natural products. In order to produce natural flavour of mushroom for the food industry, submerged fermentation of mushroom biomass, produces the "mushroom alcohol", 1-octen-3-ol, from linoleic acid by a group of enzymes including lipoxygenase and hydroperoxide lyase (Hadar and Dosoretz 1991). Many aliphatic, aromatic and terpenoid aldehydes exhibit distinct organoleptic qualities which render them valuable as flavourants in food (Stofberg and Grundschober 1987). Whole cells of the yeast *Pichia pastoris* as well as the purified enzyme, alcohol oxidase, have been used to convert various alcohols to aldehydes in aqueous and non-aqueous reaction fluids. Alcohol oxidase could be firmly immobilized on DEAE biogel to enhance stability and biocatalytic recovery (Duff and Murray 1990).

2.6 Fungicidic Activity of Aldehydes

Acetaldehyde is also known for its fungicidal and insecticidal properties. Several workers have demonstrated the ability of AA to inhibit the development of rots on various post-harvest crops. Thus, significant reduction of *Botrytis cinerea* and

Rhizopus stolonifer on strawberries and grapes and of *Penicillium expansum* on apples has been achieved by AA treatment (Prasad and Stadelbacher 1974; Stadelbacher and Prasad 1974; Morris et al. 1979; Avissar and Pesis 1991). Fumigation with other aldehydes, such as propanal and butanal has also proved effective in controlling decay in sweet cherry (Mattheis and Roberts 1993).

The fungicidic effect can be a direct effect or may be achieved through the augmentation of fruit resistance via synthesis of antifungal material. It has been reported that, in oranges, AA is the precursor, via acetyl CoA, of mevalonic acid, which is the precursor of all other monoterpenes, including the phytoalexin, limonene (Eskin 1979). Application of AA or an anaerobic treatment of N_2 or CO_2 for 24 h, which increased limonene levels, protected orange fruits against fungal decay (Pesis and Avissar 1988). The antifungal compound 1-acetoxy-2-hydroxy-5-oxo-12, 15-heneicosa-diene is produced in avocado fruits after application of 30% CO_2 for 24 h, prior to storage (Prusky et al. 1993); it increases fruit resistance to further pathogen attack.

2.7 Disadvantages of Exogenous Application of Metabolites

Exposure of plants to AA may result in adverse effects. High concentrations of AA on strawberries have caused off-flavour and injury to the caps (Prasad and Stadelbacher 1974; Aharoni et al. 1980; Pesis and Avissar 1990). Fumigation with AA has caused browning in lettuce leaves (Stewart et al. 1980). In Sultanina grape and tomato fruit, AA penetrates via the stem and causes rind damage around the stem if applied at high concentration (Pesis and Frenkel 1989; Pesis and Marinansky 1993). Also in sweet cherry, stem browning and fruit phytotoxicity have been induced by AA (Mattheis and Roberts 1993). Perata and Alpi (1991) showed AA to be the toxic compound in ethanol-induced injuries in carrot cells.

3 Induction of Aroma Volatile Production by Pre-Storage Under Anaerobic Conditions

3.1 Application of Anaerobiosis

Anaerobic conditions can be imposed in various ways:

A) Application of a high-N_2 atmosphere maybe performed by flushing the containers until most of O_2 has been removed. At this stage there are three ways to deal with the treated fruits.

1. Leave the containers closed for a short period (24–48 h); under low-O_2 atmosphere the fruits will accumulate CO_2, AA and ethanol by anaerobic respiration (Pesis and Ben Arie 1984).

2. Flush continuously with N_2 for the desired period, to flush out the CO_2, AA and ethanol.
3. The CO_2 evolved by the fruit during N_2 treatment can be scrubbed with soda lime within the container (Ben Arie et al. 1993).

B) Instead of creating anaerobic conditions by flushing with 100% N_2, it is possible either to flush the containers with 100% CO_2, to remove the air or to use different N_2/CO_2 mixtures, e.g. 49%N_2/50%CO_2/1%O_2 (Pesis and Avissar 1989; Pesis et al. 1991).

C) Creation of anaerobic conditions by enclosing the fruits in polyethylene (PE) bags under vacuum. In PE bags containing six persimmon fruits/bag, the O_2 level remained low (3%) for several weeks while the CO_2 level reached 30% after 2 days at 20 °C, in addition to the accumulated AA and ethanol (Pesis et al. 1986). It is possible to flush N_2 or CO_2 through the PE bags that had been vacuum-treated, in order to create various atmospheric conditions (Pesis et al. 1988).

D) Soaking the fruit in lime water for several days is an old method for creating the desired anaerobic conditions for removal of astringency in persimmon (Gazit and Levy 1963).

A low-O_2 atmosphere produced by N_2 or CO_2 treatments has been found to affect the fruits differently. In orange and persimmon, application of a CO_2 atmosphere for 24 h caused higher volatile production than an N_2 atmosphere (Pesis and Ben Arie 1984; Pesis and Avissar 1989; Shaw et al. 1991). The higher the CO_2 level applied, the greater the AA formation (Fig. 2).

In strawberry treated with elevated CO_2 levels (20–80% CO_2 in air), the production of AA and ethanol increased as the CO_2 level increased, and were much higher than the levels found under low O_2 (in N_2) (Ke et al. 1991). Also in tomato treated with 50–90% CO_2 (in air), as more CO_2 was applied, so more AA and ethanol were produced (Pesis and Marinansky 1993).

The increased formation of AA by CO_2-treated fruits is probably due to dark CO_2 fixation that can contribute to higher volatile production via fixation of CO_2 to malic acid by malic enzyme (Pesis and Ben Arie 1986). Plant tissue can easily metabolize applied malate or pyruvate to produce AA (Clijsters 1965; Pesis and Ben Arie 1986). In addition, it has been shown that under an anaerobic CO_2 atmosphere there is a stimulation of pyruvate decaboxylase (PDC) activity in sweet potato roots and the levels of PDC are higher than under an N_2 atmosphere (Chang et al. 1983).

3.2 Induction of Volatile Formation

Application of a short period of anaerobic conditions, such as CO_2 or N_2 atmosphere, for 24 h, prior to storage, can change fruit constituents during ripening in air. Every kind of fruit has its own sensitivity to low-O_2 conditions, e.g. apples and pears can stand low-O_2 storage for long periods (months), while subtropical

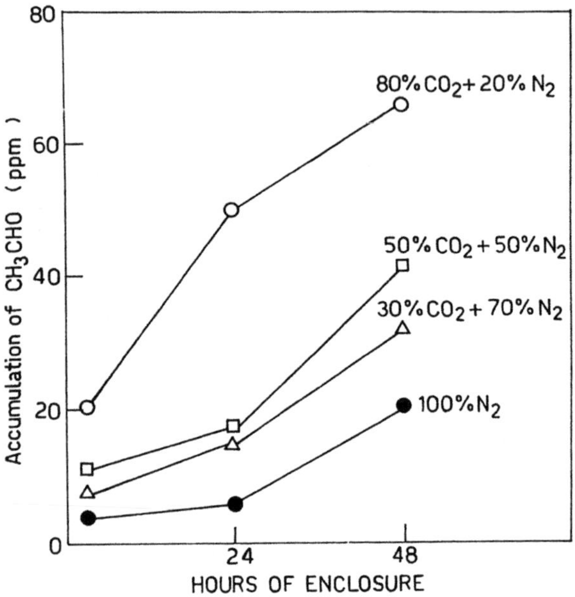

Fig. 2. Accumulation of AA (CH_3CHO) in the head space above Triumph persimmons during 48-h application of four different anaerobic atmospheres

fruits, including citrus, can stand anaerobiosis for only very short periods (1–3 days).

Usually, in commercial practice, apples are stored for extended periods under a refrigerated controlled atmosphere (CA) with O_2 reduced to 1–3%. Additional benefit for apple quality has been achieved through the application of extremely low-O_2 or high-CO_2 atmosphere, prior to cold or CA storage at 0 °C. This treatment delayed softening and ripening of several apple cultivars (Couey and Olsen 1975; Bramlage et al. 1977; Lau and Looney 1978; Little et al. 1982; Stow 1988).

Induction of AA and ethanol has been found in low-O_2 storage of apples, in which the levels of the anaerobic metabolites produced at 0% O_2 were the highest, with the level of ethanol being 100 times higher than that of AA (Patterson and Nichols 1988). Mattheis et al. (1991) showed that Delicious apples stored in 0.05% O_2/0.2% CO_2 for 30 days developed large concentrations of AA and ethanol, which led to synthesis of more fruit volatiles, mainly ethyl acetate among other ethyl esters. They showed a decrease in butyl and hexyl esters emanating from fruit previously stored under anaerobic conditions, probably because of competitive acyl esterification reactions which use the same carboxylic acids (Mattheis et al. 1991).

Pre-treatment of apples with high CO_2 can induce respiration and ethylene production at 20 °C storage, in addition to induction of other volatiles. These CO_2-treated apples were rated by panelists as better flavoured than control fruit, with the 24-h-treated fruit being the most acceptable after 2 weeks at 20 °C (Pesis et al. 1994).

Induction of volatiles following short (24–72 h) high-CO_2 or low-O_2 treatments is similar in various fruits including citrus (Norman 1977; Shaw et al. 1991), apple (Pesis et al. 1994), strawberry (Ke et al. 1991) and feijoa (Pesis et al. 1991). However, the pre-treatment did not improve the taste in every kind of fruit. In oranges, increased production of AA, ethanol, ethyl acetate and ethyl butyrate in the juice, induced by an N_2 or CO_2 atmosphere, did not cause consistent flavour changes in the juice (Shaw et al. 1991).

During storage in air, there is a decrease in the volatiles produced in various fruits, after the first volatile induction caused by the treatment. Norman (1977) showed that Navel oranges produced large amounts of AA, methanol, ethanol and ethyl acetate when stored under N_2 for 2–5 days but after removal to air, there was a gradual decrease in the production of volatiles. Progressive reduction in volatiles with increasing time at 20 °C may be a result of volatilization and/or further metabolism (Cossins 1978; De Pooter et al. 1983, 1987; Mattheis et al. 1991).

Post-harvest applicaton of various anaerobic conditions under N_2 or CO_2 for 24 h caused the accumulation of volatiles in hand-picked feijoa, significantly increasing fruit aroma and flavour. The volatiles, AA, ethanol, ethyl acetate and ethyl butyrate, remained at a high concentration for 13 days in storage at 20 °C in the treated fruits, while in the control fruits the levels were very low. A sensory panel rated the treated fruits as sweeter than the controls (Fig. 3), although there was no significant difference in the total soluble solids (TSS) or acidity between the treated and the non-treated feijoas. Treatment of feijoa fruits with 98% N_2 for 24 h was found to be the most effective for inducing volatile production (Pesis et al. 1991). The explanation for the increase in taste of the N_2-treated feijoa is that the relative contributions of ethyl acetate and ethyl butyrate to the aroma of N_2-treated feijoas after 7 days at 20 °C were much higher than those of AA or ethanol. In the non-treated fruits, the highest relative flavour contribtuion was that of the volatile AA (Pesis 1994).

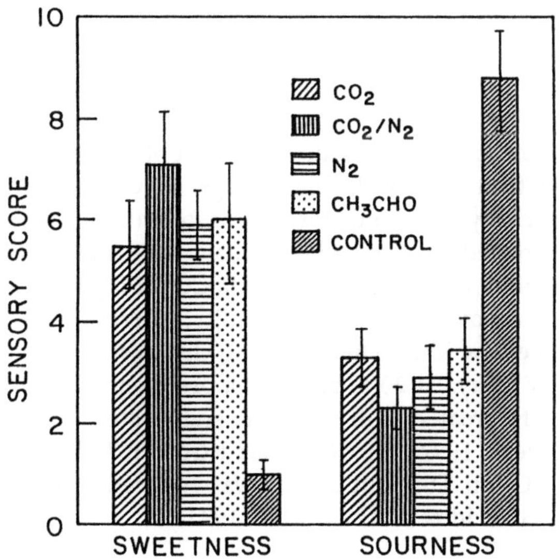

Fig. 3. Effects of anaerobic conditions and AA (CH_3CHO) application for 24 h on sensory quality of feijoa fruits 1 day after treatment at 20 °C. On the ten-point sensory quality scale, 10 = sweetest or sourest

3.3 Removal of Astringency

Astringency is a common phenomenon in many unripe fruits, but generally it disappears when the fruit ripens on the tree. There are some exceptions to this, including many varieties of persimmon and date and some of banana, that remain astringent even when horticulturally mature. Astringency is a result of high contents of soluble tannins in the pulp. Removal of astringency is necessary for the fruit to be edible.

There are several ways to remove astringency, based on using anaerobic conditions (Gazit and Levy 1963). The most commonly used method is the creation of an 80% CO_2-enriched atmosphere for 1–2 days (Matsuo and Ito 1977). Also, application of 20% ethanol vapour to astringent persimmon enclosed in polyethylene bags removed astringency in fruit still attached to the tree (Sugiura et al. 1975). In banana cv. Amas, post-harvest application of ethanol vapour (25–75%) removed astringency as measured according to total phenolic content (Esguerra et al. 1992). Sugiura et al. (1979) postulated that production of ethanol and AA in the seeds and its accumulation in the flesh is responsible for the degree of astringency found in the fruit.

It is believed that AA formed in the fruit under anaerobic or ethanol treatments is involved in both natural and artificial de-astringency processes. Matsuo and Ito (1982) demonstrated that AA reacts with kaki-tannin in vitro to form an insoluble gel. This supported the hypothesis that the soluble tannin of astringent fruit is polymerized by AA to form an insoluble compound which is non-astringent. The rate of the deastringency process was positively correlated with the level of endogenous AA attained in the fruit under CO_2 or N_2 atmosphere (Pesis and Ben Arie 1984). The higher levels of AA found in the CO_2-treated fruit than in the N_2-treated fruit can be explained through fixation of atmospheric CO_2 into malate, which leads to AA production (Pesis and Ben Arie 1986).

3.4 Changes in Sugar and Acidity

In both non-climacteric (oranges) and climacteric fruit (peach, apple) high-CO_2 pre-treatment caused reduction in fruit acidity during storage. In CO_2-treated Valencia oranges, the reduction in acidity after cold storage was in both titratable acidity and ascorbic acid, while N_2-treated and control fruits have the same levels (Pesis and Avissar 1989). In peaches and nectarines treated with CO_2 or N_2 for 24 h before storage, the CO_2-treated fruits showed a decline in acidity, whereas in the N_2-treated fruits there was no decline in acidity. There were no changes in TSS in the treated peaches. Taste tests after 5 days at 20 °C revealed a preference for the treated over the control fruit, with N_2-treated fruit being generally rated higher than CO_2-treated fruit (Lurie and Pesis 1992). In CO_2-treated Golden Delicious apples there was no influence of CO_2 on TSS, but acidity was reduced in CO_2-treated fruit, resulting in a higher TSS/acid ratio (Pesis et al. 1994).

3.5 Injury and Off-Flavour Production

When O_2 is lowered below the Pasteur point, anaerobic metabolism induces accumulation of AA and ethanol, which can lead to development of off-flavours (Fidler and North 1971).

The anaerobic conditions should be applied carefully, depending on the kind and variety of the fruit. The subtropical fruits are among the most sensitive to anaerobiosis damage. Oranges held longer than 20 h in N_2 at 20 °C exhibited rind injury (Norman 1977). An insecticidal O_2 (0.4%) atmosphere has been used in papaya for up to 3 days at 20 °C without the risk of significant fruit injury. However, after 3 days in low O_2 plus 3 days in air, the fruits developed off-flavours. The intolerance of the fruit to low O_2 was correlated with the PDC and lactate dehydrogenase activities but not with the ADH activity (Yahia et al. 1992). Different cultivars of the same fruit behave differently. Mango cv. Carabao has been reported as being sensitive to low-O_2 treatments, which cause internal breakdown (Lizada 1992), whereas mango cv. Keitt was found to be tolerant of low-O_2 or high-CO_2 atmospheres (Yahia 1993).

Strawberries have been treated with 0.25–1.0% O_2 or with 20% CO_2 in air at 0 or 5°C for 10 days without detrimental effects on quality. However, off-flavour was found in strawberries kept in 0 or 0.25% O_2; it was correlated with the concentrations of ethanol, ethyl acetate and acetaldehyde in the juice (Ke et al. 1991).

Severe off-odours, mainly compounds which contain sulphur, are produced by broccoli when it is held under anaerobic conditions which can develop after 2–3 days in modified atmosphere (MA) packages. The first compound formed under anaerobic conditions, and the one which appears to be primarily responsible for the objectionable odour of the broccoli, is methanethiol (Forney et al. 1991).

4 Conclusion

Exposure to various metabolites (aldehydes, alcohols and acids) which are precursors of aroma volatiles can lead to induction of aroma and flavour during storage of various fruits, and to improvement of their taste quality. The biosynthesis pathway of volatile esters is via acetyl CoA or acyl CoA intermediates.

The same induction of aroma volatiles can be stimulated by short-term application of anaerobic conditions (0.05–1.5% O_2) or high-CO_2 atmosphere (50–95%) in air, for a limited period prior to storage. During these treatments, there is a high induction of the anaerobic metabolites, AA and ethanol, which leads to further production of volatiles and other molecules. These treatments stimulated organoleptic attributes associated with the production of volatiles in the fruits during storage. The enhancement of fruit taste was correlated mainly with the increase in aroma volatiles. However, in some fruits, changes in sugar content or sugar/acid ratio, in the case of AA stimulation of glucenogenesis or acid degradation, could also explain the enhancement of the sensory quality.

It is necessary to emphasize that every fruit has its own sensitivity to the treatment duration or the applied concentrations. Too much exposure to anaerobic conditions or high concentrations of applied metabolites will cause damage, while too little would not be effective. For each commodity, it is necessary to study the optimal concentration or duration which will give the best result.

References

Aharoni Y, Stewart JK, Guadagni DG, Mon TR (1980) Thrips mortality and strawberry quality after vacuum fumigation with acetaldehyde or ethyl formate. J Am Soc Hortic Sci 105:926–929

Ahmed EM, Dennison RA, Dougherty RH, Shaw PE (1978) Flavour and odour thresholds in water of selected orange juice components. J Agric Food Chem 26:187–191

Avissar I, Pesis E (1991) The control of postharvest decay in table grape using acetaldehyde vapors. Ann Appl Biol 118:229–237

Baldwin EA, Nisperos-Carriedo MO, Moshonas MG (1991) Quantitative analysis of flavour and other volatiles and for certain constituents of two tomato cultivars during ripening. J Am Soc Hortic Sci 116:265–269

Bartley IM, Stoker PG, Martin ADE, Hatfield SG, Knee M (1985) Synthesis of aroma compounds by apples supplied with alcohols and methyl esters of fatty acids. J Sci Food Agric 36:567–574

Ben-Arie R, Levine A, Sonego L, Zutkhi Y (1993) Differential effects of CO_2 at low and high O_2 on the storage quality of two apples cultivars. Acta Hortic 326:165–174

Berger RG, Drawert F (1984) Changes in the composition of volatiles by postharvest application of alcohols to Red Delicious apples. J Sci Food Agric 35:1318–1325

Biale JB, Young RE (1981) Respiration and ripening in fruits-retrospect and prospect. In: Friend J, Rhodes MJC (eds) Recent advances in the biochemistry of fruits and vegetables. Academic Press, London, pp 1–39

Bramlage WJ, Bareford PH, Blanpied GD, Dewey DH, Taylor S, Porritt SW, Lougheed EC, Smith WH, McNicholas FS (1977) Carbon dioxide treatment for McIntosh apples before CA storage. J Am Soc Hortic Sci 102:658–662

Chang LA, Hammett LK, Pharr DM (1983) Carbon dioxide effects on ethanol production, pyruvate decarboxylase and alcohol dehydrogenase activities in anaerobic sweet potato roots. Plant Physiol 71:59–62

Clijsters H (1965) Malic acid metabolism and initiation of the internal breakdown in Jonathan apples. Physiol Plant 18:85–93

Cossins EA (1978) Ethanol metabolism in plants. In: Hook DD, Crawford RMM (eds) Plant life in anaerobic environments. Science Publishers, Ann Arbor, MI, pp 169–202

Couey HM, Olsen KL (1975) Storage response of Golden Delicious apples after high carbon dioxide treatment. J Am Soc Hortic Sci 100:148–150

De Pooter HL, Dirinck PJ, Willaert GA, Schamp NM (1981) Metabolism of propionic acid by Golden Delicious apples. Phytochemistry 20:2135–2138

De Pooter HL, Montens JP, Willaert GA, Dirinck PJ, Schamp NM (1983) Treatment of Golden Delicious apples with aldehydes and carboxylic acids: effect on the headspace composition. J Agric Food Chem 31:813–818

De Pooter HL, Van Acker MR, Schamp NM (1987) Aldehyde metabolism and the aroma quality of stored Golden Delicious apples. Phytochemistry 26:89–92

Dilley DR (1970) Enzymes. In: Hulme AC (ed) The biochemistry of fruits and their products, vol 1. Academic Press, New York, pp 179–207

Duff SJB, Murray WD (1990) Non aqueous reaction systems for the oxidation of higher molecular weight alcohols by alcohol oxidase from *Pichia pastoris*. In: Charalambous G (ed) Flavours and off-flavours '89. Developments in food science, vol 24. Elsevier, Amsterdam, pp 701–713

Esguerra EB, Kawada K, Kitagawa H (1992) Removal of astringency in Amas Banana (Musa AA group) with postharvest ethanol treatment. Acta Hortic 321:811–820

Eskin NAM (1979) Terpenoides and flavonoides. In: Plant pigments, flavours and textures. Academic Press, New York, pp 65–93

Fellman JK, Mattinson DS, Bostick BC, Mattheis JP, Patterson ME (1993) Ester biosynthesis in Rome apples subjected to low-oxygen atmospheres. Postharvest Biol and Technol 3:201–214

Fidler PC (1968) The metabolism of acetaldehyde by plant tissues. J Exp Bot 19:41–51

Fidler JC, North CJ (1971) The effect of periods of anaerobiosis on the storage of apples. J Hortic Sci 46:221–231

Forney CF, Mattheis JP, Austin RK (1991) Volatile compounds produced by broccoli under anaerobic conditions. J Agric Food Chem 39:2257–2259

Gazit S, Levy Y (1963) Astringency and its removal in persimmons. Isr J Agric Res 13:125–132

Gholap AS, Bandyopadhyay C, Nadkarni GB (1986) Aroma development in mango fruit. J Food Biochem 10:217–229

Gilliver PJ, Nursten HE (1976) The source of the acyl moiety in the biosynthesis of volatile banana esters. J Sci Food Agric 27:152–158

Gortner WA, Dull GG, Kraus BH (1967) Fruit development, maturation, ripening, and senescence: biochemical basis for horticultural terminology. Hortic Sci 2:141–144

Hadar Y, Dosoretz CG (1991) Mushroom mycelium as a potential source of food flavour. Trends Food Sci Technol 2:214–218

Halinska A, Frenkel C (1991) Acetaldehyde stimulation of net gluconeogenic carbon movement from applied malic acid in tomato fruit pericarp tissue. Plant Physiol 95:954–960

Hirai J, Hirata N, Horiuchi S (1968) Effect of oleification on hastening the maturity of the fig fruit. VI. Respiration and changes in concentrations of metabolic substances in the treated fruits with products in oxidative process of fatty acid such as acetaldehyde or ethylene. J Jpn Soc Hortic Sci 37:20–29

Janes HW, Chin C, Frenkel C (1978) Respiratory upsurge in blueberries and strawberries as influenced by ethylene and acetaldehyde. Bot Gaz 139:50–52

Ke DY, Goldstein L, O'Mahony M, Kader AA (1991) Effects of short-term exposure to low O_2 and high CO_2 atmospheres on quality attributes of strawberries. J Food Sci 56:50–54

Kelly MO, Saltveit ME (1988) Effect of endogenously synthesized and exogenously applied ethanol on tomato fruit ripening. Plant Physiol 88:143–147

Lau OL, Looney NE (1978) Effects of pre-storage high CO_2 treatment on British Columbia and Washington State Golden Delicious apples. J Am Soc Hortic Sci 103:341–344

Little CR, Faragher JD, Taylor HJ (1982) Effects of initial oxygen stress treatments in low oxygen modified atmosphere storage of Granny Smith apples. J Am Soc Hortic Sci 107:320–323

Lizada MC (1992) Anaerobiosis in the Carabao mango. IVth Int Mango Symp. Miami, FL Abstr 104 pp

Lurie S, Pesis E (1992) Effect of acetaldehyde and anaerobiosis as postharvest treatments on the quality of peaches and nectarines. Postharvest Biol Technol 1:317–326

Macku C, Jennings WG (1987) Production of volatiles by ripening bananas. J Agric Food Chem 35:845–848

Matsuo T, Ito S (1977) On mechanisms of removing aastringency in persimmon fruits by carbon dioxide treatment I. Some properties of the two processes in the deastringency. Plant Cell Physiol 18:17–25

Matsuo T, Ito S (1982) A model experiment for deastringency of persimmon fruit with the high carbon dioxide treatment: in vitro gelation of kaki-tanin by reacting with acetaldehyde. Agric Biol Chem 46:683–689

Mattheis JP, Roberts RG (1993) Fumigation of sweet cherry (*Prunus avium* Bing) fruit with low molecular weight aldehydes for postharvest decay control. Plant Dis 77:810–814

Mattheis JP, Buchanan DA, Fellman JK (1991) Change in apple fruit volatiles after storage in atmospheres inducing anaerobic metabolism. J Agric Food Chem 39:1602–1605

Morris JR, Cawthon DL, Buescher RW (1979) Effect of acetaldehyde on postharvest quality of mechanically harvested strawberries for processing. J Am Soc Hortic Sci 104:262–264
Nichols WC, Patterson ME (1987) Ethanol accumulation and poststorage quality of Delicious apples during short-term, low-O2, CA storage. HortScience 22:89–92
Norman SM (1977) The role of volatiles in storage of citrus fruits. Proc Int Soc Citric 1:238–242
Nunez Selles AJ, Maarse H, Bemelmans JMH (1986) Flavour changes in gamma irradiated grapefruit. Food Chem 21:183–193
Nursten HE (1970) Volatile compounds: the aroma of fruits. In: Hulme AC (ed) The biochemistry of fruits and their products, vol 1. Academic Press, New York, pp 239–268
Patterson ME, Nichols WC (1988) Metabolic response of Delicious apples to carbon dioxide in anoxic and low-oxygen environments. HortScience 23:866–868
Paz O, Janes H, Prevost B, Frenkel C (1982) Enhancement of fruit sensory quality by postharvest application of acetaldehyde and ethanol. J Food Sci 47:270–273, 276
Pereta P, Alpi A (1991) Ethanol-induced injuries to carrot cells. Plant Physiol 95:748–752
Pesis E (1994) Enhancement of fruit aroma and quality by acetaldehyde or anaerobic treatments before storage. Acta Hortic 368:365–373
Pesis E, Avissar I (1988) Effect of acetaldehyde vapors or anaerobic conditions prior to storage on postharvest quality of citrus fruits. Proc 6th Int Citrus Conf 3:1393–1399
Pesis E, Avissar I (1989) The postharvest quality of orange fruits as affected by prestorage treatments with acetaldehyde vapors or anaerobic conditions. J Hortic Sci 64:107–113
Pesis E, Avissar I (1990) Effect of postharvest application of acetaldehyde vapour on strawberry decay, taste, and certain volatiles. J Sci Food Agric 52:377–385
Pesis E, Ben-Arie R (1984) Involvement of acetaldehyde and ethanol accumulation during induced deastringency of persimmon fruits. J Food Sci 49:896–899
Pesis E, Ben-Arie R (1986) Carbon dioxide assimilation during postharvest removal of astringency from persimmon fruits. Physiol Plant 67:644–648
Pesis E, Frenkel C (1989) Effects of acetaldehyde vapors on postharvest quality of table grapes. HortScience 24:315–317
Pesis E, Marinansky R (1992) Carbon dioxide and ethylene production in response to acetaldehyde and ethanol treatments is grapes. J Am Soc Hortic Sci 117:110–113
Pesis E, Marinansky R (1993) Inhibition of tomato ripeing by acetaldehyde vapour or anaerobic conditions prior to storage. J Plant Physiol 142:717–721
Pesis E, Levi A, Ben-Arie R (1986) Deastringency of persimmon fruits by creating a modified atmosphere in polyethylene bags. J Food Sci 51:1041–1017
Pesis E, Levi A, Ben-Arie R (1988) The role of acetaldehyde production in the removal of astringency from persimmon fruits under various modified atmospheres. J Food Sci 53:153–156
Pesis E, Zauberman G, Avissar I (1991) Induction of certain aroma volatiles in feijoa by postharvest application of acetaldehyde or anaerobic conditions. J Sci Food Agric 54:329–337
Pesis E, Ampunpong C, Shusiri B, Hewett EW (1994) Induction of ethylene and CO_2 production in apple fruit following short term exposure to high CO_2. Postharvest Biol Technol 4:309–317
Pino J, Torricella R, Orsi F (1986) Correlation between sensory and gas chromatographic measurements on grapefruit juice volatiles. Nahrung 30:783–790
Poole MA (1983) Effect of postharvest applications of acetaldehyde and ethylene on the quality attributes of tomatoes. PhD Thesis, Rutgers University, New Brunswick, NJ
Prasad K, Stadelbacher GJ (1974) Effect of acetaldehyde on postharvest decay and market quality of fresh strawberries. Phytopathology 64:948–951
Prusky D, Kobiler I, Ardi R, Fishman Y (1993) Induction of resistance of avocado fruit to *Colletotrichum gloeosporioides* attack by CO_2 treatment. Acta Hortic 343:325–330
Romani RJ, Ku-Lim L (1966) Direct gas chromatographic analysis of volatiles produced by ripening pears. J Food Sci 31:558–560
Sawamura M, Tsuji T, Kuwahara S (1989) Changes in the volatile constituents of Pummelo during storage. Agric Biol Chem 53:243–246
Shaw GJ, Ellingham PJ, Birch EJ (1983) Volatile constituents of feijoa headspace analysis of intact fruit. J Sci Food Agric 34:743–747

Shaw PE, Moshonas MG, Pesis E (1991) Changes during storage of oranges pretreated with nitrogen, carbon dioxide and acetaldehyde in air. J Food Sci 56:469–474

Siek TJ, Albin IA, Sather LA, Lindsay RC (1969) Taste thresholds of butter volatiles in deodourized butteroil medium. J Food Sci 34:265–267

Stadelbacher GJ, Prasad K (1974) Postharvest decay control of apple by acetaldehyde vapor. J Am Soc Hortic Sci 99:364–368

Stewart JK, Aharoni Y, Hastell PL, Young DK (1980) Symptoms of acetaldehyde injury on head lettuce. Hortscience 15:148–149

Stofberg J, Grundschober F (1987) Consumption ratio and food predominance of flavouring materials. Perfumer Flavourist 12:27–56

Stow J (1988) The effect of high carbon dioxide pre-treatments and ethylene removal on the storage performance of apples Cox's Orange Pippin. Sci Hortic 35:89–97

Sugiura A, Harada H, Tomana T (1975) Studies on the removability of astringency in Japanese persimmon fruits I. "On tree removal" of astringency by ethanol treatment (Part I). J Jpn Soc Hortic Sci 44:265–272

Sugiura A, Yonemori K, Harada H, Tomana T (1979) Changes in ethanol and acetaldehyde contents in Japanese persimmon fruits and their relation to natural deastringency. Stud Inst Hortic Kyoto Univ 9:41–47

Teranishi R, Corse JW, McFadden WH, Black DR, Morgan AI (1963) Volatiles from strawberries. I. Mass spectral identification of the more volatile components. J Food Sci 28:478–483

Tressl R, Jennings WG (1972) Production of volatile compounds in the ripening banana. J Agric Food Chem 20:189–192

Ueda Y, Tsuda A, Bai JH, Fujishita N, Chachin K (1992) Charateristic pattern of aroma ester formation from banana, melon, and strawberry with reference to the substrate specificity of ester synthetase and alcohol contents in pulp. Nippon Shokuhin Kogyo Gakkaishi 39:183–187

Yahia EM (1993) Responses of some tropical fruits to insecticidal atmospheres. Acta Hortic 343:371–376

Yahia EM, Rivera M, Hernandez O (1992) Responses of papaya to short-term insecticidal oxygen atmosphere. J Am Soc Hortic Sci 117:96–99

Yamashita I, Nemoto Y, Yoshikawa S (1975) Formation of volatile esters in strawberries. Agric Biol Chem 39:2303–2307

Yamashita I, Nemoto Y, Yoshikawa S (1976) Formation of volatile alcohols and esters from aldehydes in strawberries. Phytochemistry 15:1633–1637

Yamashita I, Naito S, Lino K, Yoshikawa S (1978) Formation of aldehydes and alcohols from α-keto acids in strawberries. Nippon Shokuhin Kogyo Gakkaishi 25:378–382

The Analysis of Flavouring Compounds in Grapes

P.J. Williams and M.S. Allen

1 Introduction

A major stimulus to research on grape flavour has been wine production, because grape varietal flavour is a primary determinant of wine style, quality and price. This emphasis on winegrape varietal flavour contrasts with the flavour requirements of grapes destined for drying or the table market, for which the concentrations of sugars and acids are of prime importance. The flavour of dried grapes is also strongly influenced by Maillard chemistry, and because this topic has been reviewed extensively (Vernin and Párkányi 1982; Waller and Feather 1983) it will not be treated in the present work. The specific flavour attributes of table grapes have attracted little attention (Webb 1981) although floral characteristics would doubtless be appreciated and information about winegrape flavour is of relevance to this topic.

Methods for the study of grape flavour have often been employed in flavour analysis of other fruits and beverages. Techniques such as solvent extraction (Cacho et al. 1992) or headspace methods (Kallio 1991) may be used to isolate volatiles, and analysis usually involves gas-liquid chromatography (Cronin 1982), coupled with mass spectrometry (Merritt and Robertson 1982). Such methods have provided qualitative and some quantitative assessment of grape aroma composition, usually as part of research on wine flavour (Strauss et al. 1986; Rapp 1988; Etiévant 1991).

In winemaking, the fruit is predominantly *Vitis vinifera*, and varietal differences separate grapes used in the production of various wine styles. Furthermore, soil, climate and cultural practices can have an impact on flavour expression. Accordingly, a primary goal of contemporary grape flavour research is the development of methods that are capable of detecting small differences in composition. Two developments are the subject of this chapter. One is the identification and quantification of methoxypyrazines in grapes. The second is the identification in grapes of non-volatile, glycosylated conjugates of volatile molecules. These conjugates are generally present at higher concentration than the free aroma molecules and act as a source of flavour through facile acid- or enzyme-catalyzed hydrolysis, hence they are termed flavour precursors. For methoxypyrazines, a detailed discussion of the quantitative analysis of this single class of volatile compounds is appropriate. In contrast, the great diversity of compound types involved in glycosidic flavour precursors requires a broader, and rather less in-depth, coverage of methodology.

Modern Methods of Plant Analysis, Vol. 18
Fruit Analysis
Edited by H.-F. Linskens and J.F. Jackson

2 Methoxypyrazines

2.1 Isolation and Characterization

Many pyrazines have pronounced odours that contribute to food flavours. Methoxypyrazines such as 2-methoxy-3-(2-methylpropyl) pyrazine (**1**) have biosynthetic origins. They show limited structural diversity, display exceedingly low sensory detection thresholds, and contribute to the flavour of many raw vegetables and some grape varieties. In origin, they are quite different from the alkylpyrazines, such as (**2**). The latter appear to be formed by non-enzymatic reactions during roasting and cooking processes and, consequently, show greater structural diversity and contribute to the flavour of many heated food products (Fig. 1).

Identification of alkylpyrazines and methoxypyrazines proceeded rapidly in the late 1960s and early 1970s with the development of gas chromatography-mass spectrometry (GC-MS) techniques (Maga and Sizer 1973; Maga 1982). Initially, studies focussed on vegetable products, particularly bell peppers (Buttery et al. 1969a,b; Seifert et al. 1970), peas (Murray et al. 1970), potatoes (Buttery and Ling 1973; Nursten and Sheen 1974) and other vegetables (Murray and Whitfield 1975). Structures were identified, confirmed by synthesis, and their extremely low sensory detection thresholds established. In most cases, their trace levels made identification a difficult task. Three methoxypyrazines in particular were important: (**1**), (**3**) and (**4**).

Methoxypyrazine (**1**) was first identified in grapes in 1975 (Bayonove et al. 1975). Its characteristic aroma was identified during gas chromatography (GC) of an extract of Cabernet Sauvignon grapes. Structural confirmation was provided by agreement of the aroma retention time with that of authentic (**1**) on five columns of differing polarity, and identification of characteristic fragment peaks in an electron ionization (EI) mass spectrum. Later, a basic fraction

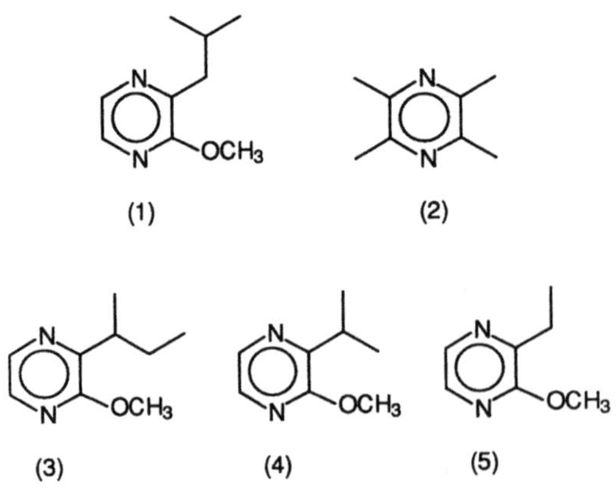

Fig. 1. Some methoxypyrazines and an alkylpyrazine identified in foods and beverages

from freon extracts of Sauvignon Blanc grapes provided evidence for the occurrence of (1), (5) and, tentatively, (4). These were identified by GC-MS using selected ion monitoring (SIM) of the important fragment ions in positive ion EI mode (Augustyn et al. 1982). These seminal studies confirmed that methoxypyrazine concentrations were very low but could not provide quantitative data. Indeed their concentration in grapes and wine was to prove far lower than the 5 µg/l speculated by Bayonove et al. (1975) and below the 600 ng/l detection limit of a HPLC method (Heymann et al. 1986). Analysis of these trace components in complex sample matrices has required GC-MS instrumentation, particularly in SIM mode. However, instrumentation is not the sole answer to the analytical difficulty; it is also important to recognize that component losses during isolation must be minimized. Perhaps this is why (1) was not found in one detailed GC-MS study of a Cabernet Sauvignon wine (Slingsby et al. 1980).

In direct contrast to the monoterpenes and norisoprenoid grape flavour components, no evidence has yet been found for the existence of bound or precursor forms of methoxypyrazines in grapes and, in particular, there is no evidence of glycosidically bound forms.

2.2 Quantitative Analysis

2.2.1 Principles of Trace Organic Analysis

Where the methodology has been developed, the most rapid and simple techniques for routine analysis of trace materials are immunoassay methods (Robins 1986). However, the techniques do not possess intrinsic reliability. Where precision and accuracy are critical, MS techniques allow a closer approach to a definitive method (Giovannini et al. 1991). At trace levels, many factors can interfere with precision and accuracy (Millard 1982). Difficulties include losses in sampling, extraction, evaporation and derivatization; adsorption on glass; variable injection volumes; decomposition during chromatography; adsorption to active sites on the column; and variation of instrumental operating conditions. To varying degrees, these may be overcome by careful choice of an internal standard added prior to isolation of the analyte from the sample matrix (Giovannini et al. 1991; Boyd 1993). The uniqueness of mass spectrometry is that it allows use of an isotopically labelled analogue of the analyte as internal standard, the technique of isotope dilution-mass spectrometry (ID-MS). As an internal standard, a stable isotope-labelled analogue of the analyte is widely recognized to be preferable to a close homologue, and the latter is better than a non-homologous compound of the same class. Benefits of a stable isotope-labelled analogue internal standard are (1) there is no possibility of its occurring naturally in any of the samples, a situation that would interfere with quantitative rigour; (2) its retention time is very close to that of the analyte, reducing the opportunity for variation of such instrumental parameters as ionization yield;

and (3) it behaves almost identically to the analyte, compensating for any analyte loss during isolation and analysis. However, a drawback to the use of a stable isotope-labelled analogue is the cost, both financially and in time, of its synthesis. Interferences from natural isotopic abundances of ^{2}H, ^{13}C and ^{18}O are best minimized by use of a standard with three or more mass units greater than the analyte; for this reason, tri-deuterated analogues are widely used. In the synthetic isotope-labelled analogue, the stability of the isotopic label in the molecule and the extent of contamination by unlabelled material must also be considered.

Quantitative analysis requires comparison of measured analyte and internal standard peak area ratios in the sample with ratios determined over a relevant concentration range with calibration standards (Giovannini et al. 1991;Boyd 1993). Analytical sensitivity can be improved and selectivity gained by using SIM to monitor only one or a few ions of the analyte and internal standard. Chemical ionization (CI) frequently increases selectivity by concentrating the ion mass profile into $(M+H)^{+}$ ions, reducing the likelihood of intrusion by ions of other sample components. Careful choice of reagent gas proton affinity can also provide selectivity of ionization (Milne and Lacey 1974). As the concentration of analyte in the sample decreases, problems of intrusion of other components into the signals of interest increase. This places increased demands upon selectivity of component isolation and chromatographic resolution, and requires verification of quantitation under different chromatographic and/or ionization conditions. In trace analysis, effective transfer of sample to the chromatography system becomes crucial to avoid molecular discrimination and to ensure that the analyte is not lost during injection.

2.2.2 Quantitative Methoxypyrazine Analysis

Determination of methoxypyrazine concentration by ID-MS was first achieved in 1987 (Harris et al. 1987) using 2-methoxy-$^{2}H_{3}$-3-(2-methylpropyl) pyrazine (**6**) (see Fig. 2) as internal standard. A Sauvignon Blanc wine was found to contain 35 ($\pm$ 2) ng/l of (**1**), ca. 6 ng/l of (**4**) and less than 0.5 ng/l of (**3**). Quantitation of (**4**) was subsequently improved by the combined use of two trideuterated internal standards (**6**) and (**7**). Application of this technique to a wide range of Sauvignon Blanc grapes and wines clearly established (**1**) as the dominant methoxypyrazine, with a range of concentrations of 0.6–38.1 ng/l across 22 wines of Australian, New Zealand and French origin (Lacey et al. 1991). An analysis of a Bordeaux wine without ID-MS techniques has reported the occurrence of (**1**) at 500 ($\pm$ 70) ng/l (Boison and Tomlinson 1990). However, the techniques used were unsuited to trace quantitative analysis, and examination of a range of Australian, New Zealand and French red wines by ID-MS, including Bordeaux wines, has established that the concentration of (**1**) is typically 50 times lower than their estimation (M.S. Allen, unpubl. results).

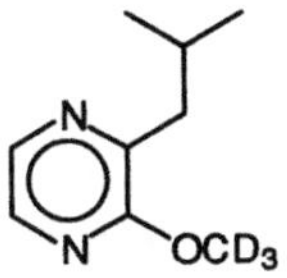

(6)

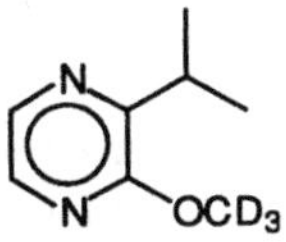

(7)

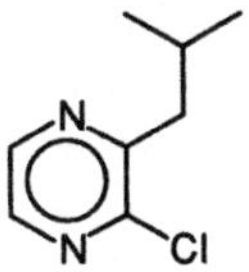

(8)

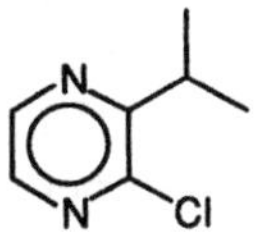

(9)

Fig. 2. Trideuterated methoxypyrazines, used as internal standards in quantitative wine analysis by isotope-dilution mass spectrometry, and the corresponding chloropyrazines used in their synthesis

2.3 Methods

2.3.1 Methoxypyrazine Standards

Unlabelled 2-methoxy-3-alkylpyrazines are commercially available (Pyrazine Specialities, Inc., Atlanta, GA, USA). They can also be synthesized by the method of Murray and Whitfield (1975). The chloro compounds (**8**) and (**9**), prepared by the method of Murray and Whitfield (1975), are converted to trideuterated methoxypyrazines (**6**) and (**7**) by the action of sodium methoxide-$^2H_3$ (Fig. 2; Harris et al. 1987; Lacey et al. 1991).

2.3.2 Instrument Calibration

To determine whether the response of the mass spectrometer is linear, it is calibrated by analysis of mixtures of (**1**) and (**6**), and mixtures of (**4**) and (**7**), at a minimum of six mole ratios covering the useful concentration range, preparing each calibration mixture in duplicate. The instrumental response (i.e. weighted linear regression of $^2H_0/^2H_3$ peak area ratio on $^2H_0/^2H_3$ mole ratio) is calculated using the reciprocal of the variance as weighting (Miller 1991). The variance can be estimated from analysis of at least six replicates at each of high and low levels of undeuterated material. Limits of detection are determined as intercept plus three standard deviations. If a linear relationship is not obtained, a number of approaches may be used to circumvent the problem (Giovannini et al. 1991).

2.3.3 Isolation of Methoxypyrazine Components

Appropriate quantities of the trideuterated methoxypyrazines (**6**) and (**7**) in purified dichloromethane are added to samples (typically 240 ml) and allowed to stand for several hours. After adjustment to pH 5, the sample is distilled at atmospheric pressure. Methoxypyrazines in the distillate (50 ml) are trapped with a strongly acidic ion-exchange resin (AG50WX4, 200 mg) then, after filtration and washing of the resin, are removed by water adjusted to pH 10 (1 ml) which in turn is extracted with dichloromethane (2 × 0.5 ml). The dichloromethane extracts are slowly concentrated under nitrogen in the dark at room temperature over 6–7 h to ca. 20 µl in tapered micro-vials, then transferred to glass capillaries that are sealed and stored in the dark at −20 °C until mass spectrometry.

2.3.4 Determination of Standard Response

Day-to-day variation of instrumental performance can be expected. It is not practical to perform a full instrumental calibration on each day of analysis, but it is necessary to compare the instrumental response with the sample to that with a standard (e.g. 1 µl of a dichloromethane solution of (**1**), (**4**), (**6**) and (**7**) at 200 µg/l, and (**3**) at 20 µg/l) on the day of analysis after, as required, mass calibration and optimization of source parameters. Quantitative analysis should be attempted only if the response to the standard is within 10% of the mean of previous satisfactory determinations.

2.3.5 Mass Spectrometry of Methoxypyrazine Isolates

Chromatography is performed with bonded phase silica capillary columns (e.g. J&W, DB-1, 60 m × 0.32 mm i.d., 1.0 µm film), using helium as carrier gas, a cool on-column inlet system, and a deactivated fused silica retention gap (e.g. J&W, 3 m × 0.32 mm i.d.) to allow injection of comparatively large solvent volumes directly onto the column. Methoxypyrazines are monitored in positive-ion CI mode using SIM (Fig. 3) at *m/z* 170.1, 167.1, 156.1 and 153.1 using ammonia (anhydrous) as reagent gas. With this arrangement, multiple injections can be made within a few minutes onto the cold column to avoid the flooded zone exceeding the retention gap and to allow the mass spectrometer vacuum system to cope with the volume of injected solvent (Lacey and Sanders 1992).

2.4 Sensory Characteristics

Methoxypyrazines can contribute to wine flavour even at low parts per trillion levels. Sensory detection thresholds in water have been determined as 2 ng/l (parts

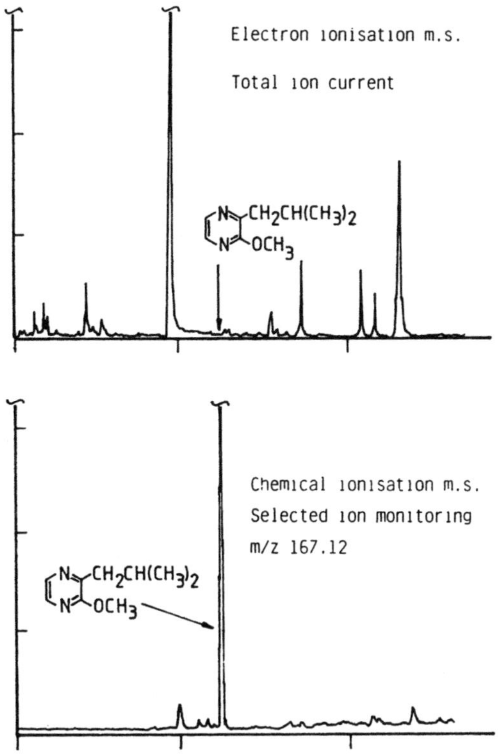

Fig. 3. Sensitivity and selectivity gain in GC-MS analysis of methoxypyazines with positive ion CI and SIM operation. The data show the determination of natural (**1**) isolated from a wine in which its concentration was ca. 30 ng/l

per trillion) for (**1**) (Buttery et al. 1969a,b; Seifert et al. 1970), 1 or 2 ng/l for (**4**) (Murray et al. 1970; Seifert et al. 1970) and 1 ng/l for (**3**) (Murray et al. 1970). While the aroma of (**1**) is characteristic of bell peppers, the aroma of (**3**) and (**4**) is more earthy, pea and potato-like. In wine, (**1**) evokes dusty, herbaceous, vegetative and green aromas, while (**4**) contributes a more earthy, boiled vegetable aroma. When added to a methoxypyrazine-free white wine, (**1**) and (**4**) were significantly detectable by duo-trio tests at 1 and 2 ng/l respectively, and (**1**) contributed recognizable vegetative aroma at 8 ng/l. In addition, commercial Sauvignon Blanc wines with high (>26 ng/l) or low (<12 ng/l) natural levels of (**1**) were clearly distinguished on the basis of sensory scores for vegetative varietal aroma (Allen et al. 1991). It has been suggested that the perceptible aroma arising from (**1**) may be modified by the low concentration of (**4**) in Sauvignon Blanc wines (Allen et al. 1988), a sensory study has indicated that (**4**) contributed more strongly to wine aroma than (**1**) when each was added to a red wine (Maga 1989). In wines, concentrations of (**1**) in the range 10–20 ng/l often contribute beneficially to flavour; however, wines with ca. 30 ng/l or more are sometimes perceived to be dominated by its aroma.

2.5 Viticultural Influences

Cool grape ripening conditions lead to an enhanced intensity of herbaceous/vegetative aroma in Sauvignon Blanc and Cabernet Sauvignon wines and a four to eightfold increase in Sauvignon Blanc grape methoxypyrazine concentrations. Since the berry methoxypyrazine concentration is comparatively high at véraison (i.e. the onset of sugar accumulation) the climatic effect may result from a lower rate of fall of methoxypyrazine concentration, under cool conditions, that is not offset by a corresponding decrease in the rate of sugar accumulation (Lacey et al. 1991). Under hot ripening conditions, the level at normal harvesting maturity may be only 1% of that present 6 weeks earlier (Allen et al. 1990a). Vine management can also influence grape methoxypyrazine levels. A comparison of pruning methods had shown that berries from minimally pruned vines had ca. eightfold lower methoxypyrazine levels than those of berries from spur-pruned vines (Allen et al. 1990b). One contributing factor is the ease of light penetration through the leaf canopy, with deeply shaded fruit having threefold higher levels than those of exposed fruit on the same vines (Allen 1993). This impact of climate, ripening and vine management on methoxypyrazine flavour levels provides a stimulus for further research into the mechanisms of methoxypyrazine control in the developing grape berry.

2.6 Varietal Differences

Most work on methoxypyrazines has focussed on the evidently herbaceous/vegetative varieties of Sauvignon Blanc and Cabernet Sauvignon. Semillon can also produce herbaceous/vegetative wines, and the occurrence of methoxypyrazines in this variety has been confirmed (Allen and Lacey 1993). Very high levels (100–200 ng/l) have been found (Calo et al. 1991) in a study of the variety Carmenere, although the authors express some reservations about possible coeluting components that may not be distinguished under the EI conditions used. A comparison of a wide range of varieties, at véraison as well as at normal maturity, has found no detectable methoxypyrazines in the varieties Pinot Noir, Malbec and Traminer, traces in Riesling and Chardonnay, and significant levels in Merlot and Cabernet Franc (Allen and Lacey 1993), confirming that methoxypyrazine occurrence is highly variety dependent.

3 Glycosidically Bound Flavour Compounds of Grapes

Initial research in this field, centring on monoterpene glycosides, helped in elucidating the role of monoterpenes as flavour compounds of floral grape varieties (Strauss et al. 1986). The additional involvement of glycosides of C_{13} norisoprenoid

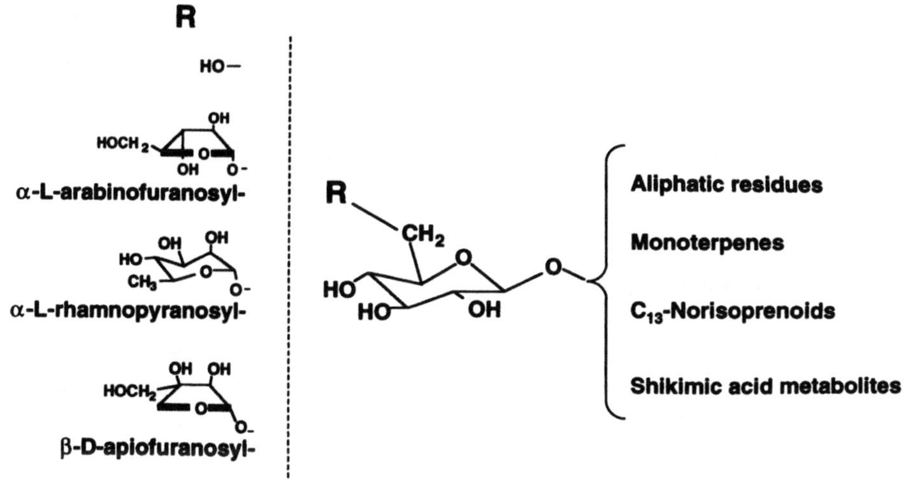

Fig. 4. Generalized structure of glycosidically bound flavour components of grapes

compounds and of shikimic acid-derived metabolites as precursors of non-floral grape flavour, was a later development (Williams et al. 1989).

Several recent reviews give, inter alia, coverage of analytical techniques that have been used in studies of glycosidic flavour precursors of fruits and plant-derived foods (Stahl-Biskup et al. 1993; Williams 1993; Williams et al. 1993b). The following discussion will emphasize the most recently published data on the analysis of flavour precursors in grapes and wines. Also covered will be sensory analytical data, the application of flavour precursor analysis in characterizing grapes and wines, and the effects on wine flavour of precursor hydrolysis.

Figure 4 shows in generalized form the structural features of those glycosidically conjugated flavour precursors that have been identified in grapes to date.

3.1 Isolation of Glycosidic Precursors

Glycoconjugates of grape juices and wines are selectively retained on C-18 reversed-phase (C-18 RP) silica gel (Williams et al. 1982) or on XAD-2 resin (Gunata et al. 1985a). Precursors can be obtained in concentrated form, free of protein, sugars, juice acids and other grossly interfering substances. The XAD resin method has been used to isolate both free and glycosidically bound monoterpenes from grapes by differential elution with pentane and ethyl acetate (Gunata et al. 1985a).

On a weight basis, C-18 RP silica gel has been shown to be the more highly efficient adsorbent (Williams 1993) and, for routine analyses, the commercial availability of cartridge packs of this material has been an advantage. The efficiency of isolation is of great importance to the reliability of analytical data obtained on grape glycosides, either for total quantification (see Sect. 3.4) or to determine the concentration of individual glycosides or aglycones for the purposes of, e.g. varietal characterization (Voirin et al. 1992b). The fidelity of such data will be flawed if the isolation method does not recover all glycosides with equal efficiency.

3.2 Analysis of Intact Glycosides

Compilations of spectrometric data for glycosides are of value in structural analysis of these compounds (Voirin et al. 1990, 1992a,b). However, the highly complex nature of grape glycosidic isolates, which are typically made up of a large number of structurally diverse aglycones in combination with several different glycosyl moieties, means that separation techniques are crucial to successful analysis. Therefore, much attention has been directed to the chromatographic separation of glycosides.

3.2.1 Gas Chromatography

Voirin et al. (1992a) have developed the use of GC and GC-MS for the analysis of derivatized grape glycosides. Trimethylsilyl (TMS) ethers and trifluoroacetate (TFA) esters of monoterpene glucosides, rutinosides, arabinoglucosides and apiosylglucosides of several grape varieties were quantified in this way (Voirin et al. 1992b). TFA derivatives were more suitable for the analysis of the higher molecular weight disaccharide conjugates, although TMS derivatives could provide complementary information.

3.2.2 High Performance Liquid Chromatography (HPLC) and Other Liquid-Solid Chromatographic Methods

Thin layer chromatography and reversed phase HPLC have found common use, both as analytical and preparative procedures, in research on grape glycosides (Marinos 1992). Combinations of liquid-solid chromatographic methods, e.g. gel chromatography, with separations based on molecular size and hydrophobic interaction, silica gel liquid partition, and over pressure liquid chromatography, have been employed in a three-step process for the separation and partial analysis of Muscat of Alexandria grape glycosides (Salles et al. 1990). HPLC on a β-cyclodextrin phase permitted assignment of the absolute stereochemistry (i.e. 95% S-) to two disaccharide conjugates of linalool in a glycosidic isolate from the same grape variety (Salles et al. 1993).

3.2.3 Countercurrent Chromatographic Methods

Countercurrent chromatography (CC) is a support-free, liquid-liquid partition system in which two immiscible liquids act as mobile and stationary phases. Absence of a solid adsorbent eliminates the possibility of analyte decomposition on an acitve surface, assures recovery of the sample material, and means that the capacity of a CC column is much greater than, e.g. a HPLC column of the same dimensions. CC has been used in glycoside research becasue it is ideally suited to separate such highly polar, water-soluble, and labile constituents without the need for their derivatization. The techniques of CC that have recently been applied in fruit flavour precursor studies, including those of grapes and wines, have been reviewed (Winterhalter 1993).

Droplet countercurrent chromatography (DCC) was the first CC technique applied to grape juice glycosides. It allowed the recognition, and for some compounds the structural identification, of glycosidic conjugates of monoterpenes, C_{13}-norisoprenoids and phenols as flavour precursors in the varieties Rhine Riesling, Muscat of Alexandria and Chardonnay (Strauss et al. 1987). Subsequent work with DCC in combination with GC-MS allowed a two-dimensional mapping of the many glycoconjugates in Riesling wine (Winterhalter et al. 1990a).

Multilayer coil countercurrent chromatography (MLCCC), has been used on the glycosides of Riesling leaves to isolate two C_{13}-norisoprenoid glucosides in sufficient amounts and purity for structural elucidation without the need for further chromatography (Roscher and Winterhalter 1993). Further work with MLCCC yielded five more C_{13}-norisoprenoid glycosides and two shikimic acid-derived conjugates from the abundant glycosides of the leaves (Skouroumounis and Winterhalter 1994). Resulting from these studies was the observation that, in Riesling vine leaves, monosaccharides were the dominant conjugates, in contrast to the common occurrence of disaccharide conjugates in wines of the variety (Roscher and Winterhalter 1993).

3.2.4 Mass Spectrometry

Soft ionization techniques allowing the direct mass spectrometric analysis of underivatized glycosides have been employed in research on fruit flavour precursors (Salles et al. 1991). Tandem mass spectrometry (MS/MS), in addition to its use in structure elucidation, has found much practical application in mixture analysis (Startin 1987) and can be considered as a form of separation technique. The combination of MS/MS and soft ionization, i.e. fast atom bombardment (FAB) MS, with partition liquid chromatography, e.g. DCC and HPLC, has been applied to the analysis of grape juice and wine glycoconjugates (Marinos 1992; Marinos et al. 1994). Positive and negative FAB-MS experiments targeted individual constituents in the liquid chromatographic sub-fractions and gave the molecular weights of the compounds. Parent and daughter ion experiments, in both positive and negative ion modes, provided further structural information on the individual constituents. This technique allowed preliminary identification and, with NMR spectrometry, the futher characterization of three new monoterpene apiosylglucosides and a new arabinoglucoside (Marinos 1992; Marinos et al. 1994).

This research showed that the mass spectrometric experiments alone can provide reliable information on the molecular weight of the glycosides, the number of sugars involved in conjugation, the sequence of sugar attachment, as well as information about the aglycone moieties. Although the data obtained with FAB-MS and MS/MS permitted only tentative structural assignments, they were obtained on a minimal amount of sample and with minimal sample preparation time because the analyses were performed on mixtures and derivatization was unneccessary. The use of FAB MS/MS on the intact glycosides can be applied as an off-line, third dimension in other analytical regimens such as the DCC-GC-MS two-dimensional mapping procedure (see Sect. 3.2.3) for the study of fruit and wine flavour precursors.

3.3 Analysis After Hydrolysis of Glycosides

Hydrolysis of glycoside isolates provides aglycones. Separate hydrolysis by acid and by glycosidase enzyme will permit assessment of the number and type of glycosidically bound constituents in the grape. Comparison of the volatile compounds produced by the two hydrolysis techniques can give information about the nature of the conjugates of particular aglycones and the possible resistance of particular conjugates to glycosidase enzyme hydrolysis (Williams et al. 1993a). The sugars involved in the glycosidic conjugates may be determined by analysis of the carbohydrates present in the acid hydrolysis residues, either by HPLC or by GC after derivatization (Herderich and Schreier 1993).

3.3.1 Enzyme Hydrolysis

The products given by glycosidase enzymic hydrolysis of a precursor isolate depend not only upon the nature of the carbohydrate conjugating moiety but also on the specificity of the enzyme(s) and, therefore, the particular enzyme preparation used (Gunata et al. 1993; Williams 1993).

Recent studies have focussed on apiosidase activity in enzyme preparations because, although many fungal enzyme preparations have a range of glycosidase activities enabling hydrolysis of arabinoglucosides and rutinosides, apiosidase activity is uncommon. Induction of apiosidase activity in a strain of *Aspergillus niger* cultured on apiin (4′,5,7-trihydroxyflavone-7-β-D-apiosyl-β-D-2-glucoside) gave an enzyme preparation with an activity capable of hydrolyzing terpenyl apiosylglycosides from grapes (Dupin et al. 1992). This research has made available fungal enzyme preparations with a broad range of glycosidase activities allowing a more complete release of aroma compounds from grape precursors in vinification (Gunata et al. 1990).

3.3.2 Compositional Analysis of Volatile Aglycones

Data from the systematic analysis of aglycones from different grape varieties are becoming available (Versini et al. 1988; Razungles et al. 1993; Sefton et al. 1993, 1994). For example, recent comparative studies have reported the determination, quantification and biogenetic classification of the acid- and glycosidase-released volatile aglycones from glycosides of Chardonnay and Sauvignon Blanc juices obtained form grapes harvested over two consecutive vintages. The free volatiles of the juices were also examined (Sefton et al. 1993, 1994).

Analysis of the glycosidically bound volatiles of Chardonnay revealed the quantitative dominance of norisoprenoids and, to a lesser extent, shikimic acid derivatives. The observation among the aglycones in these two biogenetic classes, of individually important food flavorants, e.g., damascenone, vanillin, raspberry ketone and zingerone, along with several γ-lactones in the aliphatic group, suggested the potential importance of these compounds to the varietal flavour of Chardonnay wines.

Monoterpene glycosides in Sauvignon Blanc juices were at higher concentrations than in the Chardonnay samples but at lower concentrations than those reported in muscat varieties (Versini et al. 1993b). A significant proportion (40–50%) of the bound monoterpenes were monocyclic, indicating that cyclization of, e.g., geranyl pyrophosphate is an important step in the biosynthesis of monoterpenes in Sauvignon Blanc. These monocyclic monoterpenes have limited sensory impact compared to many acyclic monoterpenes, and thus the transformations giving cyclic derivatives can be thought of as processes that "mute" the sensory contribution of monoterpenes to the variety (Williams et al. 1993a). Nevertheless, the relatively high proportion of monoterpenes found in the precursor hydrolysates of Sauvignon Blanc suggest tha this grape can be classified as a variety belonging to an intermediate class between the monoterpene-dependent floral grapes, and monoterpene-deficient non-floral fruits.

3.3.3 Sensory Analysis of Aglycones

The sensory analysis of products releasd by hydrolysis of grape glycosides is a most important aspect of precursor research, yet the subject is relatively neglected in comparison to the work that has been devoted to the chemical analysis of both aglycones and glycosides.

Some informal data indicate that wines treated enzymically to hydrolyze glycosides were more intense and complex than control samples (Gunata et al. 1993). However, formal sensory panel studies are necessary to quantify and describe the characteristics that grape glycoside hydrolysis can contribute to varietal aromas. Such formal sensory research uses descriptive analysis to provide a quantitative profile of the individual aroma attributes of a wine (Noble 1988). For good descriptive analysis the judges must be well trained and the wines must show significant differences; the latter can be assessed by, e.g., duo-trio testing (Noble 1988).

The aroma properties of hydrolyzed glycosides from Semillon, Sauvignon Blanc and Chardonnay juices have been investigated using sensory descriptive analysis. Commercial wines were available for two of the varieties of the study, i.e. Chardonnay and Semillon, and descriptive analysis of these demonstrated that some aroma attributes characterizing the wines were also aromas perceived in the hydrolyzed precursors. Thus, it was inferred that the grape precursors, on hydrolysis, contribute to and enhance key aromas of the wines (Francis et al. 1992).

A study of the black grape Shiraz used descriptive analysis to evaluated volatile components liberated from the glycosidic precursors by hydrolysis at pH 3.2 or by non-selective glycosidase enzymic hydrolysis (Abbott et al. 1991). The research was given an extra dimension by inclusion of Shiraz juices assigned as high and low quality, each obtained from two regions over 2 years, as well as wines made from the juices of one of the years. The results showed that the quality differences could be distinguished and quantified by the sensory methods. The glycosidic hydrolysates prepared by both enzymic- and acid hydrolysis were found to contribute non-berry attributes to wines such as stalky, earthy and cigar-tobacco aromas. Importantly, it was found that wines considered to be of high quality were rated higher in non-berry attributes than their low quality counterparts. From the work it was deduced that glycosidic hydrolysates from Shiraz grapes contain aroma compounds that are important to high quality wines of the variety.

Such research formally links the aroma characteristics of hydrolyzed grape flavour precursors and wines, and thus provides a sensory justification for the chemical analysis of the glycosidic fraction of grapes. The data also emphasize the importance of acid hydrolysis to the liberation of sensory-significant products from grape glycosidic precursor fractions. Of the grapes that have been studied to date, acid hydrolysis was the only technique that gave material detectable by the panel for all four varieties.

3.4 Application of Flavour Precursor and Aglycone Compositional Analysis Data to Grape and Wine Characterization

Early work in this field studied the occurrence and quantification of several monoterpene diols released enzymatically from glycosidic fractions of grapes of different varieties (Strauss et al. 1988). This approach has been extended with quantification of individual free and bound compounds, mostly monoterpenes, in a range of muscat and other floral grapes (Voirin et al. 1992b; Rapp et al. 1993; Versini et al. 1993a,b). The variability inherent in such data has been demonstrated, particularly by the work of Versini et al. (1993a), although statistical treatment of sufficiently large data sets has permitted these researchers to discriminate among many muscat varieties (Versini et al. 1993b).

Differences in total concentration of distinct terpene classes, i.e. the monoterpenes and norisoprenoids, that were released from the glycosidic precursors of the varieties Chardonnay and Sauvignon Blanc, were discussed in Section 3.3.2. It is evident that with extension of this approach to other non-floral varieties (Versini et al. 1988; Razungles et al. 1993) the reliability of distinguishing compositional data is likely to be further established.

Another early research aim was to follow the evolution of free and glycosidically bound flavour compounds in growing fruit (Wilson et al. 1984; Gunata et al. 1985b) and much current work in this area still places emphasis on monoterpene compounds in floral varieties (Park et al. 1991; Razungles et al. 1993; Versini et al. 1993a). These studies are essential to establish a data base of compositional variability arising from the effects of grape ripening. Variables of this nature, together with the wide concentration ranges of glycosidically bound monoterpenes observed in grapes of the same variety and analyzed at approximately the same level of maturity (Versini et al. 1993a), must be assessed before good use can be made of flavour precursor analysis in appraising viticultural influences on grape composition and flavour.

Viticultural researchers seek quick and convenient methods for analysis of grape flavour compounds, suitable for processing the large number of samples that can be generated in field trials. The development of a rapid distillation technique for the estimation of free, and potentially volatile, monoterpene compounds in grapes was a first step in satisfying these demands (Dimitriadis and Williams 1984). This method has found use in studies of fruit flavour composition influenced by ripening (Marais 1987), vine microclimate (Reynolds and Wardle 1989), fruit cluster exposure (Macaulay and Morris 1993), as well as clonal and pruning effects (McCarthy 1992). It is limited to monoterpenes, however, and is thus restricted to use on floral grape varieties. This has highlighted the need for a more widely applicable technique that would allow an assessment of the concentration of total glycosidically bound flavour constitutents in all grape varieties.

The basis of such a broadly applicable assay is the recognition that all glycosylated secondary metabolites that have been identified in grapes are glucosides, with the glucose moiety usually further substituted, i.e. to give a disaccharide. Hydrolysis of grape glycosides, therefore, yields equimolar proportions of aglycones and D-glucose. Based on this, a determination of the glycosyl-glucose (G-G) concentration will permit an estimation of the total concentration of glycosylated secondary metabolites present in the fruit, i.e. the G-G might indicate the concentration, inter alia, of the difficult-to-measure flavour compounds (Abbott et al. 1993).

A method involving the following steps has now been established to measure the G-G of grape juices (Williams et al. 1995):

1. Isolation of a glycosidic fraction from the juice by selective retention of the glycosides on a C-18 RP adsorbent.

2. Acid hydrolysis of this glycosidic fraction to liberate the glucose bound in the fraction.
3. Measurement of the concentration of the released glucose using an enzymic determination specific for that compound.

It is anticipated that the availability of this method will greatly facilitate the accumulation of data on glycosidically bound secondary metabolites and the responses of these compounds to the influences of growing conditions.

3.5 Effects on Wine Aroma Resulting From Precursor Hydrolysis

Fungal-derived glycosidase enzyme preparations have been promoted in the wine industry for the aroma enhancement of wines, and some data on compositional changes resulting from the application of such commercial preparations are available (Cuénat et al. 1992).

Interest in techniques alternative to enzyme treatment for aroma enhancement has been stimulated by the data discussed in Section 3.3.3 demonstrating the wide applicability of acid hydrolysis for the release of aroma from grape glycosidic precursors. One approach appropriate to modern winemaking technology is the employment of heat to effect hydrolysis of the grape flavour precursors catalyzed by the natural acids of the wine. The sensory and compositional effects on wines that have been subjected to such processing have been reported (Leino et al. 1993; Francis et al. 1994). Head-space analysis demonstrated that heat treatment altered volatile ester and norisoprenoid compositions; the latter effects were consistent with a mechanism involving hydrolysis of glycosidic precursors.

Not all volatiles released from grape glycosides have desirable effects on wine aroma, and data on the development of compounds detrimental to wine quality, such as the hydrocarbon 1,1,6-trimethyl-1,2-dihydronaphthalene (TDN), have recently become available (Marais et al. 1992a,b,c). TDN is a thermodynamically stable end product formed by acid-catalyzed rearrangement of a multiplicity of glycosylated norisoprenoids (Winterhalter et al. 1990b). Wine storage time and temperature, as well as fruit growing conditions all appear to have an effect on the TDN concentration of Riesling wines.

4 Conclusion

This chapter has identified two major challenges in the analysis of flavour components of grapes. One is that components may be present at trace or ultra-trace levels, and this places special demands on detection, characterization and quantitative analysis. The second is that components also exist in precursor forms which, in the case of the glycosides, necessitate techniques suited to analysis of relatively involatile compounds with wide structural diversity. Mass spectrometry, coupled either directly or used off-line with a range of chromatographic techniques, has played a crucial role in both challenges. Undoubtedly, developments in mass spectrometry will continue rapidly, and such developments will shape our future analytical horizons. In addition, our knowledge of the sensory properties of individual constituents, and of the biosynthesis, transport and regulation of flavour components and flavour precursors, are likely to play increasing roles. Nevertheless, with the analysis methods currently available, research is already underway that could ultimately provide an understanding of varietal differences among grape cultivars, the development of varietal flavour in wines, and suggest techniques for the post-harvest enhancement of grape varietal flavour in wines, and for the quality evaluation of a grape crop.

References

Abbott NA, Coombe BG, Williams PJ (1991) The contribution of hydrolyzed flavor precursors to quality differences in Shiraz juice and wines: an investigation by sensory descriptive analysis. Am J Enol Vitic 42:167–174

Abbott NA, Williams PJ, Coombe BG (1993) Measure of potential wine quality by analysis of grape glycosides. In: Stockley CS, Johnstone RS, Leske PA, Lee TH (eds) Proceedings of the eighth Australian wine industry technical conference, 25–29 Oct 1992, Winetitles, Melbourne, pp 72–75

Allen MS (1993) Viticultural effects on methoxypyrazine grape flavour: current research directions. Aust Grapegrower Winemaker 354:10–12

Allen MS, Lacey MJ (1993) Methoxypyrazine grape flavour components: influence of grape cultivar. In: Stockley CS, Johnstone RS, Leske PA, Lee TH (eds) Proceedings of the eighth Australian wine industry technical conference, 25–29 Oct 1992, Winetitles, Melbourne, pp 195

Allen MS, Lacey MJ, Harris RLN, Brown WV, (1988) Sauvignon blanc varietal aroma. Aust Grapegrower Winemaker April:51,53–54,56

Allen MS, Lacey MJ, Brown WV, Harris RLN (1990a) Occurrence of methoxypyrazines in grapes of *Vitis vinifera* cv. Cabernet Sauvignon and Sauvignon blanc. In: Ribereau-Gayon, P, Lonvaud A (eds) Actualités Oenologiques 89, Comptes rendus 4e Symposium International d'Oenologie, 15–17 June 1989, Bordeaux. Dunod, Paris, pp 25–30

Allen MS, Tassie E, Lacey MJ, Brown WV, Harris RLN(1990b) Influence of pruning methods on methoxypyrazine flavour components of Cabernet Sauvignon grapes. In: Williams PJ, Davidson D, Lee TH (eds) Proceedings of the seventh Australian wine industry technical conference, 13–17 Aug 1989. Winetitles, Adelaide, pp 247

Allen MS, Lacey MJ, Harris RLN, Brown WV (1991) Contribution of methoxypyrazines to Sauvignon blanc wine aroma. Am J Enol Vitic 42:109–112

Augustyn OPH, Rapp A, van Wyk CJ (1982) Some volatile aroma components of *Vitis vinifera* L. cv. Sauvignon blanc. S Afr J Enol Vitic 3:53–60

Bayonove C, Cordonnier R, Dubois P (1975) Etude d'une fraction caractéristique de l'arôme du raisin de la variété Cabernet-Sauvignon; mise en evidence de la 2-methoxy-3-isobutylpyrazine. C R Acad Sci Paris Ser D 281:75–78

Boison JO, Tomlinson RH (1990) New sensitive method for the examination of the volatile flavour fraction of Cabernet Sauvignon wines. J Chromatogr 522:315–327

Boyd RK (1993) Quantitative trace analysis by combined chromatography and mass spectrometry using external and internal standards. Rapid Commun Mass Spectrom 7:257–271

Buttery RG, Ling LC(1973) Earthy aroma of potatoes. J Agric Food Chem 21:745–746

Buttery RG, Seifert RM. Guadagni DG, Ling LC (1969a) Characterization of some volatile constituents of bell peppers. J Agric Food Chem 17:1322–1327
Buttery RG, Seifert RM, Lundin RE, Guadagni DG, Ling LC (1969b) Characterisation of an important aroma component of bell peppers. Chem Ind (Lond) 12 April 1969:490–491
Cacho J, Melendez J, Ferreira V (1992) Development of a method for analyzing volatiles from foodstuff matrices, including microextraction by demixture. Application to the analysis of grapes. Mikrochim Acta 108:61–72
Calo A, Stefano RD, Costacurta A, Calo G (1991) Caratterizzazione di Cabernet Franc e Carmenere (*Vitis* sp.) e chiarimenti sulla loro coltura in Italia. Riv Vitic Enol 44:3–25
Cronin DA (1982) Techniques of analysis of flavours: chemical methods including sample preparation. In: Morton ID, MacLeod AJ (eds) Food flavours. Part A. Introduction. Elsevier, Amsterdam, pp 15–48
Cuénat P, Canal-Llaubères R-M, Leyat C (1992) Autoenrichissement des moûts et libération d'arômes dans les vins. Rev Suisse Vitic Arboric Hortic 24:73–74
Dimitriadis E, Williams PJ (1984) The development and use of a rapid analytical technique for estimation of free and potentially volatile monoterpene flavorants of grapes. Am J Enol Vitic 35:66–71
Dupin I, Gunata Z, Sapis J-C, Bayonove C (1992) Production of β-apiosidase by *Aspergillus niger:* partial purification, properties, and effect on terpenyl apiosylglucosides from grape. J Agric Food Chem 40:1886–1891
Etiévant PX (1991) Wine. In: Maarse H (ed) Volatile compounds in foods and beverages. Marcel Dekker, New York, pp 483–546
Francis IL, Sefton MA, Williams PJ (1992) Sensory descriptive analysis of hydrolysed precursor fractions from Semillon, Chardonnay, and Sauvignon blanc grape juices. J Sci Food Agric 59:511–520
Francis IL, Sefton MA, Williams PJ (1994) The sensory effects of pre- or post-fermentation thermal processing on Chardonnay and Semillon wines. Am J Enol Vitic 45:243–251
Giovannini MG, Pieraccini G, Moneti G (1991) Isotope dilution mass spectrometry: definitive methods and reference materials in clinical chemistry. Ann Ist Super Sanità 27:401–410
Gunata YZ, Bayonove CL, Baumes RL, Cordonnier RE (1985a) The aroma of grapes. I. Extraction and determination of free and glycosidically bound fractions of some grape aroma components. J Chromatogr 331:83–90
Gunata YZ, Bayonove CL, Baumes RL, Cordonnier RE (1985b) The aroma of grapes. Localisation and evolution of free and bound fractions of some grape aroma components cv. Muscat during first development and maturation. J Sci Food Agric 36:857–862
Gunata Z, Dugelay I, Sapis JC, Baumes R, Bayonove C (1990) Action des glycosidases exogènes au cours de la vinification: libération de l'arôme à partir de précurseurs glycosidiques. J Int Sci Vigne Vin 24:133–144
Gunata Z, Dugelay I, Sapis JC, Baumes R, Bayonove C (1993) Role of enzymes in the use of the flavour potential from grape glycosides in winemaking. In: Schreier P, Winterhalter P (eds) Progress in flavour precursor studies: analysis – generation – biotechnology. Allured Publishing, Carol Stream, pp 219–234
Harris RLN, Lacey MJ, Brown WV, Allen MS (1987) Determination of 2-methoxy-3-alkylpyrazines in wine by gas chromatography/mass spectrometry. Vitis 26:201–207
Herderich M, Schreier P (1993) High-resolution gas chromatography-mass spectral analysis of carbohydrates using methoxyoxime acetates. Phytochem Anal 4:256–260
Heymann H, Noble AC, Boulton RB (1986) Analysis of methoxypyrazines in wines. 1. Development of a quantitative procedure. J Agric Food Chem 34:268–271
Kallio H (1991) Method of sensitive analysis of wine headspace volatiles based on selective capillary column trapping. J Chromatogr Sci 29:438–443
Lacey MJ, Sanders CJ (1992) Chemical composition of sex pheremone of oriental fruit moth and rates of release by individual female moths. J Chem Ecol 18:1421–1435
Lacey MJ, Allen MS, Harris RLN, Brown WV (1991) Methoxypyrazines in Sauvignon blanc grapes and wines. Am J Enol Vitic 42:103–108
Leino M, Francis IL, Kallio H, Williams PJ (1993) Gas chromatographic headspace analysis of Chardonnay and Semillon wines after thermal processing. Z Lebensm Unters Forsch 197:29–33

Macaulay LE, Morris JR (1993) Influence of cluster exposure and winemaking processes on monoterpenes and wine olfactory evaluation of Golden Muscat. Am J Enol Vitic 44:198–204

Maga JA (1982) Pyrazines in foods: an update. CRC Crit Rev Food Sci Nutr 16:1–48

Maga JA (1989) Sensory and stability properties of added methoxypyrazines to model and authentic wines. In: Charalambous G (ed) Flavors and off-flavors, proceedings of the 6th international flavour conference, Rethymnon, Crete, Greece, 5–7 July 1989. Elsevier, Amsterdam, pp 61–70

Maga JA, Sizer CE (1973) Pyrazines in foods: a review. J Agric Food Chem 21:22–30

Marais J (1987) Terpene concentrations and wine quality of *Vitis vinifera* L. cv. Gewürztraminer as affected by grape maturity and cellar practices. Vitis 26:231–245

Marais J, van Wyk CJ, Rapp A (1992a) Effect of storage time, temperature and region on the levels of 1,1,6-trimethyl-1,2-dihydronaphthalene and other volatiles, and on quality of Weisser Riesling wines. S Afr J Enol Vitic 13:33–44

Marais J, Versini G, van Wyk CJ, Rapp A (1992b) Effect of region on free and bound monoterpene and C_{13}-norisoprenoid concentrations in Weisser Riesling wines. S Afr J Enol Vitic 13:71–77

Marais J, van Wyk CJ, Rapp A (1992c) Effect of sunlight and shade on norisoprenoid levels in maturing Weisser Riesling and Chenin blanc grapes and Weisser Riesling wines. S Afr J Enol Vitic 13:23–32

Marinos VA (1992) Application of fast atom bombardment mass spectrometry to the examination of glycoconjugates in grape juice and wine. PhD Thesis, University of Adelaide, Adelaide

Marinos VA, Tate ME, Williams PJ (1994) Protocol for FAB MS/MS characterization of terpene disaccharides of wine. J Agric Food Chem 42:2486–2492

McCarthy MG (1992) Clonal and pruning effects on Muscat à Petite Grains Blanc yield and terpene concentration. Am J Enol Vitic 43:149–152

Merritt C, Robertson DH (1982) Techniques of analysis of flavours: gas chromatography and mass spectrometry. In: Morton ID, MacLeod AJ (eds) Food flavours. Part A. Introduction. Elsevier, Amsterdam, pp 49–78

Millard BJ (1982) Sources of error in quantitative mass spectrometry. In: Facchetti S (ed) Applications of mass spectrometry to trace analysis: lectures of a course held at the Joint Research Centre, 29 Sept–3 Oct 1980, Ispra, Italy. Elsevier, Amsterdam, pp 163–179

Miller JN (1991) Basic statistical methods for analytical chemistry. Part 2. Calibration and regression methods. Analyst 116:3–14

Milne GWA, Lacey MJ (1974) Modern ionization techniques in mass spectrometry. CRC Crit Rev Anal Chem 4(1):45–104

Murray KE, Whitfield FB (1975) The occurrence of 3-alkyl-2-methoxypyrazines in raw vegetables. J Sci Food Agric 26:973–986

Murray KE, Shipton J, Whitfield FB (1970) 2-Methoxypyrazines and the flavour of green peas (*Pisum sativum*). Chem Ind (Lond) 4 July 1970:897–898

Noble AC (1988) Analysis of wine sensory properties. In: Linskens HF, Jackson JF (eds) Wine analysis. Springer, Berlin Heidelberg New York, pp 9–28

Nursten HE, Sheen MR (1974) Volatile flavour components of cooked potato. J Sci Food Agric 25:643–663

Park SK, Morrison JC, Adams DO, Noble AC (1991) Distribution of free and glycosidically bound monoterpenes in the skin and mesocarp of Muscat of Alexandria grapes during development. J Agric Food Chem 39:514–518

Rapp A (1988) Wine aroma substances from gas chromatographic analysis. In: Linskens HF, Jackson JF (eds) Wine analysis. Springer, Berlin Heidelberg New York, pp 29–66

Rapp A, Suckrau I, Versini G (1993) Untersuchungen des Trauben- und Weinaromas: Beitrag zur Sortencharakterisierung neutraler Rebsorten (Silvaner, Weißburgunder, Ruländer). Z Lebensm Unters Forsch 197:249–254

Razungles A, Gunata Z, Pinatel S, Baumes R, Bayonove C (1993) Étude quantitative de composés terpéniques, norisoprénoïdes et de leurs précurseurs dans diverses variétés de raisins. Sci Aliment 13:59–72

Reynolds AG, Wardle DA (1989) Influence of fruit microclimate on monoterpene levels of Gewürztraminer. Am J Enol Vitic 40:149–154

Robins RJ (1986) The measurement of low-molecular-weight, non-immunogenic compounds by immunoassay. In: Linkens HF, Jackson JF (eds) Immunology in plant sciences. Springer, Berlin Heidelberg New York, pp 86–141
Roscher R, Winterhalter P (1993) Application of multilayer coil countercurrent chromatography for the study of *Vitis vinifera* cv. Riesling leaf glycosides. J Agric Food Chem 41:1452–1457
Salles C, Jallageas J-C, Crouzet J (1990) Chromatographic separation and partial identification of glycosidically bound volatile components of fruit. J Chromatogr 522:255–265
Salles C, Jallageas J-C, Fournier F, Tabet J-C, Crouzet JC (1991) Apricot glycosidically bound volatile components. J Agric Food Chem 39:1979–1983
Salles C, Jallageas J-C, Crouzet JC (1993) HPLC separation of fruit diastereoisomeric monoterpenyl glycosides. J Essent Oil Res 5:381–390
Sefton MA, Francis IL, Williams PJ (1993) The volatile composition of Chardonnay juices: a study by flavor precursor analysis. Am J Enol Vitic 44:359–370
Sefton MA, Francis IL, Williams PJ (1994) Secondary metabolites of *Vitis vinifera* cv. Sauvignon blanc. J Food Sci (in press)
Seifert RM, Buttery RG, Guadagni DG, Black DR, Harris JG (1970) Synthesis of some 2-methoxy-3-alkyplyrazines with strong bell pepper-like odors. J Agric Food Chem 18:246–249
Skouroumounis GK, Winterhalter P (1994) Glycosidically bound norisoprenoids from *Vitis vinifera* cv. Riesling leaves. J Agric Food Chem 42:1068–1072
Slingsby RW, Kepner RE, Muller CJ, Webb AD (1980) Some volatile components of *Vitis vinifera* variety Cabernet Sauvignon wine. Am J Enol Vitic 31:360–363
Stahl-Biskup E, Intert F, Holthuijzen J, Stengele M, Schultz G (1993) Glycosidically bound volatiles – a review 1986–1991. Flavour Fragr J 8:61–80
Startin JR (1987) Application of mass spectrometry/mass spectrometry in food analysis. In: Gilbert J (ed) Applications of mass spectrometry in food science. Elsevier Applied Science, London, pp 289–341
Strauss CR, Wilson B, Gooley PR, Williams PJ (1986) Role of monoterpenes in grape and wine flavor. In: Parliment TH, Croteau R (eds) Biogeneration of aromas. ACS symp ser no 317. American Chemical Society, Washington, DC, pp 222–242
Strauss CR, Gooley PR, Wilson B, Williams PJ (1987) Application of droplet countercurrent chromatography to the analysis of conjugated forms of terpenoids, phenols, and other-constituents of grape juice. J Agric Food Chem 35:519–524
Strauss CR, Wilson B, Williams PJ (1988) Novel monoterpene diols and diol glycosides in *Vitis vinifera* grapes. J Agric Food Chem 36:569–573
Vernin G, Párkányi C (1982) Mechanisms of formation of heterocyclic compounds in Maillard and pyrolysis reactions. In: Vernin G (ed) Chemistry of heterocyclic compounds in flavours and aromas. Ellis Horwood, Chichester, pp 151–207
Versini G, Dalla Serra A, Dell'Eva M, Scienza A, Rapp A (1988) Evidence of some glycosidically bound new monoterpenes and norisoprenoids in grapes. In: Schreier P (eds) Bioflavor 87. de Gruyter, Berlin, pp 161–170
Versini G, Rapp A, Dalla Serra A (1993a) Considerations about the presence of free and bound *p*-menth-1-enediols in grape products. In: Schreier P, Winterhalter P (eds) Progress in flavour precursor studies: analysis – generation – biotechnology. Allured Publishing, Carol Stream, pp 243–249
Versini G, Dalla Serra A, Monetti A, De Micheli L, Mattivi F (1993b) Free and bound grape aroma profiles variability within the family of muscat-called varieties. Connaissance aromatique des cepages et qualité des vins, 1993. Montpellier, France
Voirin S, Baumes R, Bayonove C, M'Bairaroua O, Tapiero C (1990) Synthesis and n.m.r. spectral properties of grape monoterpenyl glycosides. Carbohydr Res 207:39–56
Voirin SG, Baumes RL, Gunata ZY, Bitteur SM, Bayonove CL (1992a) Analytical methods for monoterpene glycosides in grape and wine. I. XAD-2 extraction and gas chromatographic-mass spectrometric determination of synthetic glycosides. J Chromatogr 590:313–328
Voirin SG, Baumes RL, Sapis J-C, Bayonove, CL (1992b) Analytical methods for monoterpene glycosides in grape and wine. II. Qualitative and quantitative determination of monoterpene glycosides in grape. J Chromatogr 595:269–281

Waller GR, Feather MS (1983) The Maillard reaction in foods and nutrition. American Chemical Society, Washington, DC
Webb AD (1981) Quality factors in California grapes. In: Teranishi R, Barrera-Benitez H (eds) Quality of selected fruits and vegetables of North America. American Chemical Society, Washington DC, pp 1–9
Williams PJ (1993) Hydrolytic flavor release in fruit and wines through hydrolysis of non-volatile precursors. In: Acree TE, Teranishi R (eds) Flavor science, sensible principles and techniques, American Chemical Society, Washington, DC, pp 287–303
Williams PJ, Strauss CR, Wilson B, Massy-Westropp RA (1982) Use of C_{18} reversed-phase liquid chromatography for the isolation of monoterpene glycosides and nor-isoprenoid precursors from grape juice and wines. J Chromatogr 235:471–480
Williams PJ, Sefton MA, Wilson B (1989) Nonvolatile conjugates of secondary metabolites as precursors of varietal grape flavor components. In: Teranishi R, Buttery RG, Shahidi F (eds) Flavor chemistry trends and developments, 5–11 June 1988, Toronto, Ontario, Canada. American Chemical Society, Washington, DC, pp 35–48
Williams PJ, Francis IL, Sefton MA (1993a) Sensory and chemical analysis of hydrolysed flavour precursors from Sauvignon Blanc grapes. In: Schreier P, Winterhalter P (eds) Progress in flavour precursor studies: analysis – generation – biotechnology. Allured Publishing, Carol Stream, pp 235–242
Williams PJ, Sefton MA, Marinos VA (1993b) Hydrolytic flavour release from non-volatile precursors in fruits, wines and some otehr plant-derived foods. In: Hopp R, Mori K (eds) Recent developments in flavor and fragrance chemistry: proceedings of the 3rd Haarmann and Reimer international symposium, 12–15 Apr 1992, VCH, Kyoto, Japan, pp 283–290
Williams PJ, Cynkar W, Francis IL, Gray JD, Iland PG, Coombe BG (1995) Quantification of glycosides in grapes, juices, and wines through a determinationn of glycosyl glucose. J Agric Food Chem (in press)
Wilson B, Strauss CR, Williams PJ (1984) Changes in free and glycosidically bound monoterpenes in developing muscat grapes. J Agric Food Chem 32:919–924
Winterhalter P (1993) Application of countercurrent chromatographic techniques in flavour precursor studies. In: Schreier, P, Winterhalter P (eds) Progress in flavour precursor studies: analysis – generation – biotechology. Allured Publishing, Carol Stream, pp 31–44
Winterhalter P, Sefton MA, Williams PJ (1990a) Two-dimensional GC-DCCC analysis of the glycoconjugates of monoterpenes, norisoprenoids, and shikimate-derived metabolites from Riesling wine. J Agric Food Chem 38:1041–1048
Winterhalter P, Sefton MA, Williams PJ (1990b) Volatile C_{13}-norisoprenoid compounds in Riesling wine are generated from multiple precursors. Am J Enol Vitic 41:277–283

Analysis of Bitter Principles in *Citrus*

S. Hasegawa, M.A. Berhow, and C.H. Fong

1 Introduction

Bitterness is a major problem to the citrus industry. It is of economic importance because bitter juices have a lower market value. Bitterness is mainly caused by the accumulation of two different chemical compounds: limonin from the limonoid terpine group and naringin from the flavonoid phenolic group in the fruit tissues. The highest concentrations of the bitter compounds are generally found in immature fruits. As the fruits mature, the concentrations of the bitter compounds decrease.

Limonin bitterness is especially acute in juice obtained from short maturation time fruits such as navel oranges. Juices obtained from most of the other citrus crops, such as Valencia oranges, usually do not have the limonin bitterness problem, as the concentration of limonin has decreased sufficiently by the time the processing season starts. Juices from navel oranges, particularly early- to mid-season fruits, can contain over 30 ppm of limonin, which is well above the bitterness detection threshold of 6 ppm. Grapefruit also has significant levels of limonin, which vary from 10 ppm in the early season to 1 ppm or less in the late season.

Nomilin, a limonoid that is approximately twice as bitter as limonin (Hashinaga et al. 1977), was found in various citrus species. Nomilin concentrations ranging from 1.6 to less than 0.1 ppm were found in commercial grapefruit juices with the greatest concentration found in early-season juices (Rouseff 1982). The concentrations in the early season exceed the detection threshold of 0.8 ppm but they are below the recognition threshold of 2.1 ppm (Rouseff and Matthews 1984). Other bitter limonoids are also present in the citrus juices, but at much lower concentrations than limonin. They do not contribute significantly to the bitterness problem.

The flavanone neohesperidosides are found only in citrus cultivars related to the pummelo, such as grapefruit, sour orange and natsudaidai. Grapefruit is cultivated extensively throughout the world and has high levels of naringin in the fruit and juice in addition to the significant levels of limonin. It is interesting to note that pummelo has very low levels of the limonoid glucosides in its juice. Naringin levels vary from over 600 ppm in the early season grapefruit to below 100 ppm in the late season in some varieties. The threshold of detection for naringin is in the range of 20 to 50 ppm. A comprehensive study

Modern Methods of Plant Analysis, Vol. 18
Fruit Analysis
Edited by H.-F. Linskens and J.F. Jackson

on the limonin and naringin thresholds was conducted by Guadagni et al. (1973, 1974a,b).

Currently, bitter juice is dealt with in two ways commercially. The most common method is by blending. Exceptionally bitter juices are often blended with nonbitter juices to lower the naringin and/or limonin levels to levels below the threshold of detection. The other common debittering method is to pass the juice through various adsorption, ion exchange, and hollow fibre columns to remove some of the bitter compounds. Commercial debittering systems using this methodology are available. Cyclodextrins have also been used as an additive to juice as they reduce the perceived bitterness of both naringin and limonin (Shaw 1990). Naringinase enzyme preparations from fungal cultures are available which rapidly convert naringin to tasteless compounds by the hydrolysis of the sugar groups at room temperature and the acidic pHs normally found in juices, but are also considered to be an additive if mixed with juice (Hasegawa and Maier 1981). Some work has been done on the use of immobilized naringinase columns as a method of debittering juice, but the practical application of this technology has been fairly limited. A novel limonoid debittering technique was developed by Hasegawa and his coworkers in which juice is passed through a column made of immobilized cells of bacteria such as *Arthrobacter globiformis* and *Corynebacterium fascians*. This immobilized bacteria in the column converts the bitter limonin and nomilin to non-bitter forms by the action of the limonoate dehydrogenase enzyme present in the bacteria (Hasegawa and Pelton 1983). The ability to successfully transform citrus cell cultures and regenerate genetically altered citrus plants (Hidaka et al. 1990; Vardi et al. 1990) suggests alternative solutions to the bitterness problem.

2 Limonoid Bitterness

Limonoids are a group of chemically related triterpene derivatives found in the Rutaceae and Meliaceae families. There are two groups of limonoids; the aglycones and their corresponding glucosides. A total of 36 limonoid aglycones have been isolated from *Citrus* and its hybrids (Table 1). Five of them are bitter in taste - limonin, nomilin, nomilinic acid, obacunoic acid and ichangin (Fig. 1). In citrus juices, limonin is the major cause of delayed bitterness.

The observed phenomenon that a variety of citrus juices are not bitter-tasting when freshly juiced, but turn quite bitter a few hours after juicing, is known as delayed bitterness. Intact fruit tissues contain limonoate A-ring lactone (LARL), the non-bitter monolactone precursor of limonin (Maier and Beverly 1968). When the fruit is juiced, a second lactone ring gradually forms on the LARL, converting it to bitter limonin (Fig. 2). This conversion is accelerated by the presence of an endogenous enzyme, limonin D-ring lactone hydrolase, which catalyzes D-ring

Table 1. Limonoid aglycones in *Citrus* and its hybrids

Neutral	Acidic
1. Limonin	24. Deacetylnomilinic acid
2. Nomilin	25. Nomilinic acid
3. Obacunone	26. Isoobacunoic acid
4. Deacetylnomilin	27. Epiisoobacunoic acid
5. Ichangin	28. Isolimonic acid
6. Deoxylimonin	29. Deoxylimonic acid
7. Deoxylimonol	30. 17-Dehydrolimonoic acid A-ring lactone
8. Limonol	31. Calaminic acid
9. Limonyl acetate	32. Retrocalaminic acid
10. 7α-Obacunol	33. Cyclocalaminic acid
11. 7α-Obacunyl acetate	34. Isoobacunoic acid diosphenol
12. Ichangensin	35. Obacunoic acid
13. Citrusin	36. 19-Hydroxydeacetylnomilinic acid
14. Calamin	
15. Retrocalamin	
16. Cyclocalamin	
17. Methyl isoobacunoate diosphenol	
18. Methyl deacetylnomilinate	
19. 6-Keto-7β-Deacetylnomilol	
20. 6-Keto-7β-Nominol	
21. Isocyclocalamin	
22. 1-(10→19)Abeo-7α-acetoxy-10β-hydroxyisoobacunoic acid 3, 10-lactone	
23. 1-(10→19)Abeo-obacun-9(11)-en-7α-yl acetate	

lactoneization under acidic conditions, or opens the D-ring of limonin under alkaline conditions (Maier et al. 1969). It is believed that each limonoid aglycone listed in Table 1, which are generally known as dilactones has a corresponding open D-ring monolactone form. The monolactones are the predominant form of limonoid aglycones in fruit tissues while the dilactones are the predominant form in seeds.

Nomilin the general biosynthetic precursor of all limonoids, is biosynthesized via the terpenoid pathway from acetate and/or mevalonate in stems. Nomilin is then translocated to other plant tissues, including leaves, seeds, and fruit tissues, where it is converted to the other limonoids independently (Hasegawa et al. 1986; Ou et al. 1988). In most citrus species and hybrids, nomilin is converted to obacunone and then to obacunoate. Obacunaote is then converted to limonin probably via the intermediate ichangin. Nomilinic acid is the predominant acidic limonoid in citrus, but it does not seem to be directly involved in the biosynthetic pathways of major limonoids.

Hasegawa et al. (1989) found that limonoids are present as glucoside derivatives in *Citrus* tissues. Seventeen limonoid glucosides have been isolated from *Citrus* species (Table 2); each having a single D-glucose molecule linked at the C-17 position. All limonoid glucosides have been found to be tasteless. Limonoid aglycones are converted to their glucoside derivatives during late stages of

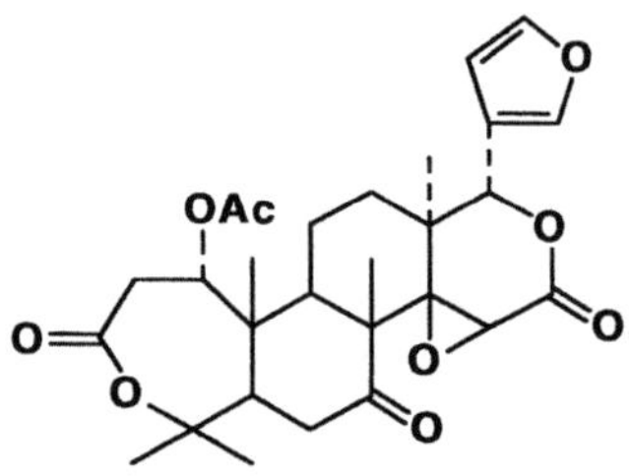

Nomilin

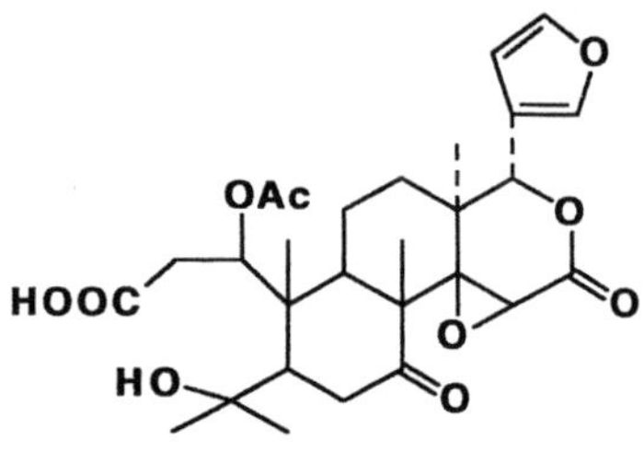

Nomilinic acid

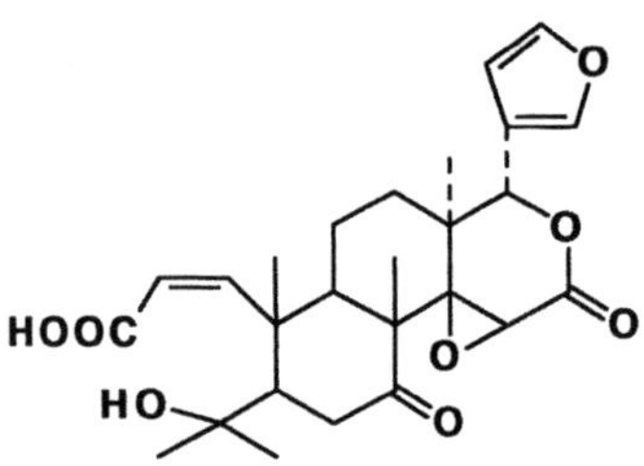

Obacunoic acid

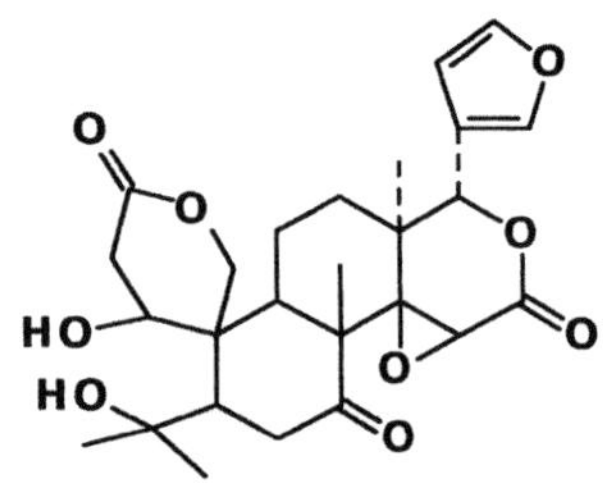

Ichangin

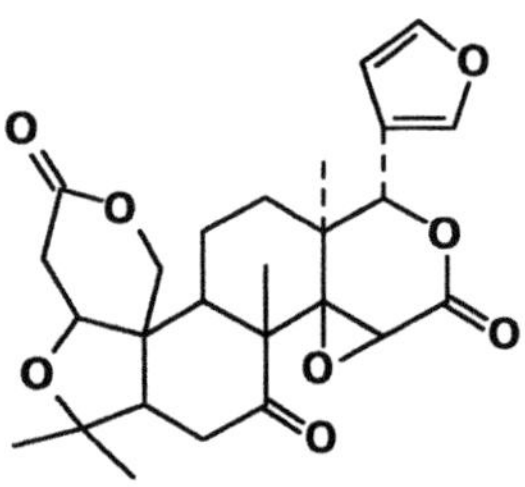

Limonin

Fig. 1. The bitter limonoids

fruit growth and maturation in fruit tissues and seeds (Hasegawa et al. 1991; Fong et al. 1992). The glucosidation of LARL to form limonin 17-β-D-glucopyranoside is the reason that limonin level in citrus juices decreases in late season. Stems and leaves are unable to glucosidate limonoids (Fong et al. 1991). For additional information on the biosynthesis of limonoids in citrus, see Hasegawa and Herman (1992).

Limonoid aglycones have been shown to possess certain biological activities such as the inhibition of cancerous tumour growth in laboratory animals (Lam et al. 1989 a,b; Miller et al. 1989) and antifeedant activities against insects (Alford and Bentley 1986; Liu et al. 1990). Recently, the glucoside of limonin, limonin 17-β-D-glucopyranoside, was found to possess anticancer activities as well (Miller et al. 1992).

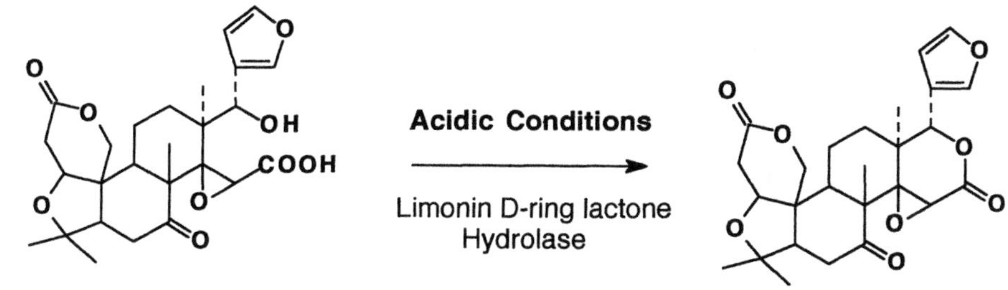

Limonin A-ring lactone (tasteless)

limonin (Bitter)

Fig. 2. The mechanism of delayed bitterness in citrus juices

Table 2. Limonoid glucosides in *Citrus*

Monocarboxylic acids	
17-β-D-glucopyranosides of:	
1. Limonin	2. Nomilin
3. Deacetylnomilin	4. Obacunone
5. Ichangin	6. Ichangensin
7. Calamin	8. 6-Keto-7β-deacetylnomilol
9. Methyl deacetylnomilinate	
Dicarboxylic acids	
17-β-D-glucopyranosides of:	
10. Nomilinic acid	11. Deacetylnomilinic acid
12. Obacunoic acid	13. *Trans*-obacunoic acid
14. Isoobacunoic acid	15. Epiisoobacunoic acid
16. Isolimonic acid	17. 19-Hydroxydeacetylnomilinic acid

2.1 Limonoid Aglycone Analysis

A variety of methods have been proposed for the detection and quantification of bitter limonoids in citrus. Most methods were developed specifically for the quantification of limonin in citrus juices. The early techniques included colorimetry, fluorimetry, and gas chromatography.

The colorimetric method (Wilson and Crutchfield 1968) and the fluorometric method (Fisher 1973) involve the chemical derivation of the limonoids. However, the reactions are not specific for limonin, and these techniques depend on isolating limonin from interfering compounds. Kruger and Colter (1972) developed a derivitization procedure to allow for the non-volatile limonin in citrus juices to be analyzed by gas chromatography.

Table 3 shows a summary of the recent techniques used and their applicability. Each of these techniques will be discussed in the following sections. Other review

Table 3. Analytical techniques for limonoids

Technique	Sample	Limonoid Quantified[a]
TLC	Tissue extracts	L, N, and others
	Juice	L, N, and others
HPLC	Tissue extracts	L, N, and others
	Juice	L, N
RIA	Tissue extracts	L
	Juice	L
EIA	Juice	L

[a]L = limonin, N = nomilin.

papers on analysis of limonoid aglycones in citrus juices and limonoids in citrus seeds have been published recently (Johnson 1988; Herman et al. 1992).

2.1.1 Sample Preparation

Some analytical techniques for the assay of limonin in citrus juices, such as radioimmunoassay and enzyme-linked immunoassay, do not require sample preparation. For HPLC and most of the TLC methods, extraction of limonoid aglycones from the juice is required. There are two different basic methods of extraction: extracting the juice directly or after mixing the juice with diatomaceous earth in a coarse-porosity fritted disc funnel and extracting via vacuum filtration with organic solvents (Kimball 1991) such as chloroform, methylene chloride and ethyl acetate; or removing the limonoids from the juice by passing it through solid phase extraction columns such as C-18 Sep-Pak cartridges.

Sample preparation for citrus tissues is different from citrus juices. Monolactones are the predominant form of limonoids in most intact citrus tissues, but most of the analytical methods can measure only the dilactone forms. To analyze the total of both forms in citrus tissues, the monolactones must first be extracted from the tissues and converted to the dilactones during sample preparation. Acidification and the use of heat are the two main methods of converting monolactones into dilactones.

Maier and Grant (1970) mixed finely macerated fruit tissue with water. The mixture was adjusted to pH 3 with HCl, butylated hydroxy-toluene added, and boiled for 5 min prior to chloroform extraction. Mansell and Weiler (1980) first extracted citrus tissues with acetone for 90 min at 55 °C. The acetone portion was then diluted with 0.01 N HCl for radioimmunoassay.

While the monolactones are soluble in aqueous solution, the dilactones have poor solubility in water. This poses a problem for extracting seeds, which have high concentrations of dilactones. Hasegawa et al. (1980) macerated citrus seeds in 0.1 M Tris buffer at pH 8.0 to convert the dilactones into monolactones by opening the D-ring. The mixture was incubated for 18 h at room temperature. After filtering, the filtrate was acidified to pH 2.0. The acidic mixture was then extracted with CH_2Cl_2. Hasegawa et al. (1984, 1991) later used a similar sample preparation procedure for HPLC and TLC analyses of other citrus tissues. The incubation period was eliminated and the tissues extracted directly by ethyl acetate with 2,6-di-tert-butyl-p-cresol added as an antioxidant. Fong et al. (1992) used a modified procedure for the extraction of Valencia orange fruit tissues. After the extraction with Tris buffer and acidification, the acidic extract was passed through a reverse-phase C-18 Sep-Pak cartridge and eluted with methanol. The methanol fraction was used for TLC analysis for various limonoids.

Rouseff and Nagy (1982) used a direct solvent extraction method for citrus seeds. Oils were removed from seed meals with hexane in a Soxhlet extractor. The seed residues were then extracted with chloroform. The hexane fraction was saved for analysis of obacunone, which is somewhat soluble in hexane.

Miyake et al. (1991) found that extraction of neutral limonoids from Natsudaidai seeds depended upon the solvents used, time of extraction, and methods of sample preparation prior to extraction. Since the seeds contain high concentrations of both dilactones and monolactones with dilactones dominating, direct organic solvent extraction method yields lower values than extraction with the Hasegawa method mentioned above. The optimum extraction method was determined to be grinding the seeds in 0.1 M Tris buffer at pH 7.0 and incubating the mixture for at least 20 h at room temperature. It was followed by acidification and extraction with methylene chloride. The organic phases can then be evaporated off to concentrate the samples for analysis.

2.1.2 Thin-Layer Chromatography

TLC is the easiest and most economical technique of detecting limonoid aglycones. Unlike other methods, TLC can be used for a variety of tissue samples. TLC solvent systems were well developed and an appropriate system can be found for a particular type of tissue sample with just a few trials. The major disadvantage is the use of subjective visual comparison with standards for qualitation and quantitation.

Maier and Grant (1970) extracted the fruit tissue-acid mixture, which was prepared as described in previous section, with chloroform. The chloroform extract was then evaporated to dryness and re-dissolved in acetonitrile. The acetonitrile sample was then spotted on silica gel TLC plates along with the appropriate standards, and developed in a benzene-ethanol-water-acetic acid solvent system (200:47:15:1, upper phase).

Tatum and Berry (1973a) spotted 50-μl aliquots of commercial juice samples directly on a 20 × 20 cm silica gel G plate (Analtech Inc.) without extraction or other sample preparation. The plate was scribed into 1-cm channels prior to the spotting; and standards of 0.1 through 0.5 μg were spotted in the centre five channels of the plate. The plate was thoroughly dried with a heat gun; and the dried plate was first developed with acetone to 3 cm mark from the applied spots. The acetone was used to extract limonin out from the dried juice solids. Then the plate was dried and developed in one of the 17 solvent systems listed in the paper.

The Hasegawa group has been quite successful with the use of only four TLC solvent systems for various tissue extracts on silica gel plates. They are: (A) ethyl acetate-cyclohexane (3:2); (B) methylene chloride-methanol (97:3); (C) methylene chloride-ethyl acetate (3:2); and (D) toluene-ethanol-H_2O-acetic acid (200:47:15:1, upper layer). Since acidic aglycones have very low R_f values in these solvent systems, they are usually methylated with diazomethane. The resulting methyl esters can then be analyzed. Each limonoid aglycone can be separated and positively identified using two or three of the above solvent systems. The R_f values for limonin, nomilinic, ichangin, and the methyl ester of nomilin acid and obacunoic acid using solvent system (A) are 0.27, 0.19, 0.12, 0.38, and 0.42, respectively. Herman et al. (1992) listed R_f values for 21 other limonoid aglycones.

Since the solvent system which separates the limonoids in one sample may not be satisfactory for another, several solvent systems should be tried for a particular sample in order to find the one that has the best separation. For example, Fong et al. (1992) used solvent system (A) to quantify limonin and nomilin in the extract of Valencia orange flesh, and solvent system (C) to quantify deacetylnomilin in the same extract. For the analysis of the extract from Valencia orange peel, system (B) was used to quantify limonin, nomilin and deacetylnomilin.

After the TLC plate is developed, there are two main methods of visualization: (1) spraying the dried plate with 10% sulphuric acid solution in ethanol, followed by heating in an oven at 125 °C or on a heating plate to develop dark-coloured spots; (2) spraying the dried plate with Ehrlich's reagent (1% p-dimethylaminobenzaldehyde in ethanol), followed by treating the plate with hydrogen chloride gas, which can be generated by mixing sulphuric acid with ammonium chloride in a hood chamber giving the limonoids a very distinctive reddish-orange colour.

Final estimation of the limonoid spots in made by visual or spectrodensitometric comparison with the standards.

2.1.3 High-Performance Liquid Chromatography

In the food industry, high-performance liquid chromatography (HPLC) is the most widely used technique for the determination of limonin levels in citrus juices, since it is accurate and reliable. Most of the HPLC methods developed are for the analysis of limonin in citrus juices. These methods may not work for other bitter limonoid aglycones and/or tissues.

A refractometer can be used for detecting limonoids if there are no other interfering compounds, but UV detector is preferred due to its greater sensitivity. Limonin has an absorption maximum at 207 nm. However, due to the significant absorption of other compounds in the extract at this wavelength, the absorption at other wavelengths such as 210, 214 or higher have been used for detection. The same wavelengths can be used for other limonoids.

Rouseff and Fisher (1980) developed a normal-phase HPLC method for analysis of limonin and nomilin in citrus juices. Limonin and nomilin were separated on a DuPont Zorbax cyano (CN) 5-μm column, 4.6 × 250 mm at 40 °C. The mobile phase consisted of a hexane-isopropanol-methanol mixture [11:12:2 (v/v)] and a flow rate of 1.0 ml/min was used for optimum resolution and sensitivity. Retention times for limonin and nomilin are 11 min and 9 min, respectively. A mixture of 6:3:1 (v/v) with a flow rate of 1.5 ml/min was recommended for routine quality control purpose. UV detectin at 207 nm was used.

Shaw and Wilson (1984) developed a more rapid analytical method using solid phase extraction with reverse phase HPLC. Juice was passed through a C-18 Sep-Pak solid phase extraction cartridge, washed with water , then eluted with acetonitrile. However, the acetonitrile extract was determined not to be stable overnight, even at −16 °C (Shaw 1986). A Brownlee C-18 5-μm microbore column, 2.1 × 220 mm, or a Perkin-Elmer C-8 5-μm column, 4.6 × 125 mm, with a 4.6 × 40 mm Brownlee 5-μm C-8 guard column was used. The mobile phase consisted of an acetonitrile-tetrahydrofuran-water mixture [17.5:17.5:65 (v/v/v)] and a flow rate of 0.2 ml/min was used for the C-18 column. A mixture of 17.5:15:67.5 with a flow rate of 1.5 ml/min was used for the C-8 column. Shaw (1986) listed methanol-acetonitrile-water (26.5:21.5:52) as the optimum mobile phase for C-18 column. This mobile phase was reported to have better baseline stability and freedom from negative peaks. UV detection at 207 nm was used.

To clean up their samples for reverse phase HPLC analysis, Van Beek and Blaakmeer (1989) investigated the use of C-18, C-8, C-2, cyclohexyl, phenyl, and CN solid phase extraction columns. No improvement of selectivity between limonin and interfering components in grapefruit juice was found. It was reported that the only difference is the elution strength of wash solvent and limonin eluent. Acetonitrile-water mixtures gave better results than methanol-water mixtures.

Widmer (1991) reported good selectivity using a CN column under reversed phase conditions for analysis of limonin in citrus juices. Juice was passed through C-18 Sep-Pak solid phase extraction cartridge, the cartridge was washed with 30% (v/v) aqueous methanol, and was eluted with 70% methanol. The 70% methanol extract was used for HPLC analysis. Unlike acetonitrile extract, the 70% methanol extract was found to be stable for 30 days if kept at −4 °C. The extract was separated on a Supelco 5-μm CN column, 4.6 × 250 mm, at 30 °C. A Brownlee 4.6 × 30 mm CN guard column was recommended. A mobile phase of acetonitrile and water (38:62) with a flow rate of 1.5 ml/min was used after washing the column with 10 ml acetonitrile. Retention time for limonin is about 6 min. UV detection at 210 nm or 214 nm was used.

The above HPLC methods are used for the analysis of limonin and/or nomilin in citrus juices. Herman et al. (1989) used a reverse phase HPLC method to quantify limonoids in fruits, seeds, leaves, stems and seedlings of *Citrus ichangensis*. The extraction procedure of Hasegawa et al. (1984) as described in Section 2.1.1 was followed. The dried extracts were dissolved in 50% methanol prior to injection. A reverse phase C-18 column (4.6 × 250 mm) was used. The column was eluted isocratically with water-methanol-acetonitrile (49:41:10) at a flow rate of 1 ml/min. Limonoids were detected by UV absorption at 210 nm. This method was used to quantify limonin, nomilin, deacetylnomilin, obacunone, ichangensin, deacetylnomilinic acid, isolimonic acid and nomilinic acid. The acidic limonoids were quantified as methyl ester derivatives.

Hasegawa et al. (1991) used a linear gradient system to analyze limonin in navel orange fruit tissues. After extracting the tissues with Tris buffer and ethyl acetate as described in Section 2.1.1, the ethyl acetate extract was evaporated to dryness and dissolved in methanol for storage at −88 °C. The methanol fraction was used for HPLC analysis. An Alltech reverse phase C-18 column, Spherisorb ODS-11, 5-μm (4.6 × 250 mm) was used. For whole fruit and peel analyses, the column was eluted with a linear gradient system starting with 10% acetonitrile in 3 mM phosphoric acid and ending with 50% acetonitrile in 40 min. Limonin eluted at 41 min. For flesh samples, the linear gradient system was the same except that the gradient was run for 60 min instead of 40 min. Limonin eluted at 55 min.

Rouseff and Nagy (1982) extracted the seeds of eight citrus species with hexane and chloroform. The chloroform extract was evaporated to dryness and redissolved in acetonitrile. The procedure of Rouseff and Fisher (1980), which was mentioned above, was followed. The mobile phase was heptane-isopropanol-methanol (1:12:2). Limonin, nomilin, obacunone and deacetylnomilin were quantified.

2.1.4 Radioimmunoassay (RIA)

Radioimmunoassay methods using antibodies prepared against limonin-linked proteins and assayed with either ^{125}I-linked tracers (Mansell and Weiler 1980) or tritium-linked tracers (Weiler and Mansell 1980) were developed specifically for the assay of limonin. RIA can be used for the quantification of limonin in various fruit tissues, seeds, leaves and twigs.

Deoxylimonin and deacetylnomilin were found to cross-react with the limonin antibody, but the relatively low levels of these two compounds in citrus tissues should not add to the limonin values significantly.

In comparison to values obtained from HPLC, the RIA limonin values from commercial grapefruit juices were found to be higher (Rouseff and Mansell 1982). It is possible that the limonoid glucosides in the juice cross-react with the limonin antibody, which results in artificially higher limonin determinations.

RIA is a very rapid technique and several hundred juice samples can be processed per day. It is also very sensitive and concentrations down to parts per billion of limonin can be detected. However, its major disadvantage is the requirement of specialized laboratories and costly counting equipment.

2.1.5 Enzyme-Linked Immunoassay (EIA)

Since many laboratories are not equipped for radiochemical analysis, an immunoassay method utilizing anti-limonin antibodies produced in rabbits and limonin-alkaline phosphate tracer was developed to measure limonin in *citrus* juices (Jourdan et al. 1984). The method was once marketed commercially and used in commercial plants. However, the method has been phased out of the food industry and is no longer available commercially.

This method can be directly used without extraction or preparation of the juice samples, several samples can be analyzed simultaneously, and it is faster than HPLC methods. However, EIA has a problem with reproducibility (Widmer and Rouseff 1991); and it is less accurate than HPLC.

Limonoid glucosides and deoxylimonin were found to cross-react with the antibodies. The relatively low levels of deoxylimonin in citrus juices should not contribute to the limonin values significantly. However, the high concentration of limonoid glucosides, which can be over 400 ppm, can give artificially high limonin values.

Kimball (1991) pointed out that besides the interference from limonoid glucosides, the failure to properly calibrate EIA test systems and the irreproducibility of results in juice samples analyzed at different Brix levels are the primary reasons for the less reliable result of EIA.

2.2 Sources of Limonoid Standards

Limonin has recently become commercially available from several chemical companies including Aldrich and Sigma. Nomilin is also available from Takasago (Toyko, Japan). Other limonoid aglycones and limonoid glucosides are currently only obtained by isolation from citrus seeds. See Herman et al. (1992) for extraction and isolation methods.

3 Flavonoid Bitterness

Flavonoid bitterness is caused by one or more of the four flavanone neohesperidosides accumulated in citrus species related to the pummelo. This includes such economically important species as the sour orange and the grapefruit. Pummelo accumulates the flavanone neohesperidoside glycosides exclusively, while sour orange and grapefruit also accumulate both flavanone neohesperidosides and the tasteless flavanone rutinosides found in other citrus species related to the citron and the mandarin (Albach and Redman 1969). The taste properties of the flavonoid glycosides are related to the sugar linkage as well as some other factors (Horowitz and Gentili 1977). The structures of the flavanone glycosides found in the *Citrus* species are shown in Fig. 3. The most common glycosidic group attached to the flavonoids present in citrus is the rhamnoglucose diglycoside. This sugar group is present in two isomeric forms which are shown in Fig. 3 – neohesperidose and rutinose. The two sugars differ only in the position of

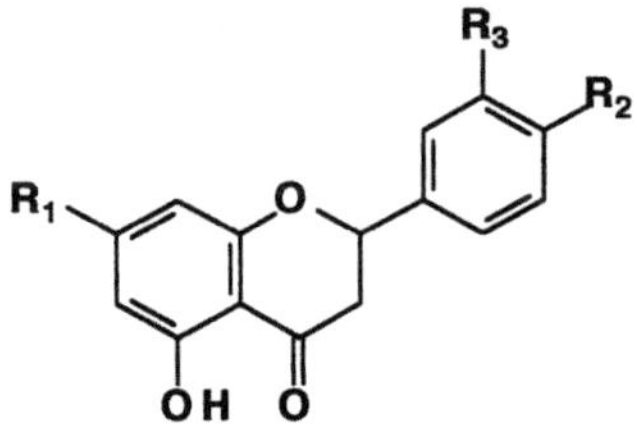

Flavanone

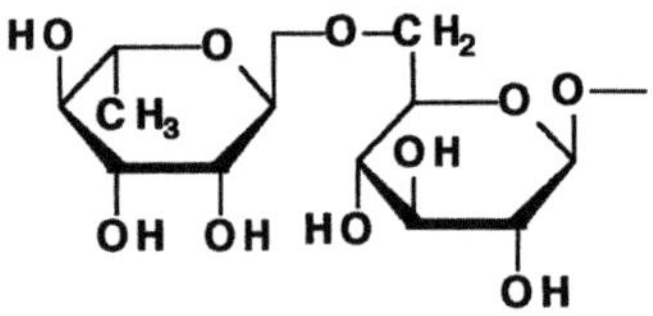

rutinoside
(6-O-α-L-rhamnosyl-β-D-glucoside)

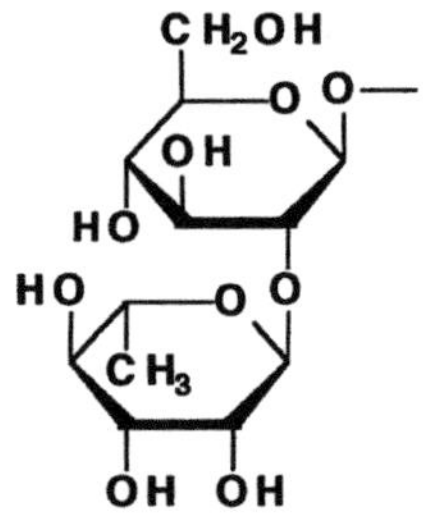

neohesperidoside
(2-O-α-L-rhamnosyl-β-D-glucoside)

Fig. 3. Structures of the major citrus flavanones. Neohesperidosides (neo.) (*bitter*) – naringin (R_1 = neo., R_2 = OH, R_3 = H), poncirin (R_1 = neo., R_2 = OMe, R_3 = H), neoeriocitrin (R_1 = neo., R_2 = OH, R_3 = OH), neohesperidin (R_1 = neo., R_2 = OMe, R_3 = OH). Rutinosides (rut.) (*tasteless*) -narirutin (R_1 = rut., R_2 = OH, R_3 = H), didymin (R_1 = rut., R_2 = OMe, R_3 = H), eriocitrin (R_1 = rut., R_2 = OH, R_3 = OH), hesperidin (R_1 = rut., R_2 = OMe, R_3 = OH)

the attachment of rhamnose to the glucose. The flavanone neohesperidosides are bitter, and the bitterness is relative to the composition of the flavanone aglycone structure. Thus, the relative bitterness of the flavonoid neohesperidosides is as follows (refer to Fig.3 for structure identification):

naringin (e) > poncirin (f) > neohesperidin (h) > neoeriocitrin (g).

Naringin is by far the most predominant flavonoid bitter principle in grapefruit (Hagen et al. 1966) and pummelo, while neohesperidin is slightly more predominant in sour orange (Castillo et al. 1992). Neoeriocitrin and poncirin occur in only relatively minor amounts in juices.

The flavanone rutinosides are tasteless, as are both the flavone neohesperidosides and flavone rutinosides (Horowitz and Gentili 1977). However, some of the flavone neohesperidosides act to raise the threshold concentration at which the bitterness of both limonin and naringin is detectd (Guadagni et al. 1973, 1974a). Opening the A ring of the flavanone neohesperidosides with alkali and a subsequent reduction yields the dihydrochalcone (DHC) neohesperidosides which are generally very sweet. Neohesperidin DHC is 1000 times sweeter than sucrose (Horowitz and Gentili 1986). The presence of the flavanone rutinosides interferes with the simple colorimetric assays for bitter flavanone neohesperidosides, requiring a more rigorous methodology to determine absolute concentrations.

The flavonoids are derived from phenylalanine via the general phenylproponoid biosynthetic pathway. The biosynthesis in *Citrus* is similar to other plant species (Hasegawa and Maier 1981). This pathway is highly regulated and controlled by both normal growth and development as well as being induced by wounding and attack by pathogens. For general reviews of flavonoid biosynthesis in plants see Heller and Forkmann (1988); Hahlbrock and Scheel (1989); Stafford (1990). In *Citrus*, the flavonoid glycosides are accumulated in young leaves and fruits during the cell division stage (Fisher 1968; Berhow and Vandercook 1989). The woody portions of the plant do not biosynthesize the characteristic flavonoids of the leaves and fruits, though these compounds are found in stem and root tissues (Jourdan et al. 1985b; Berhow and Vandercook 1991). The flavanone glycosides are apparently stored in the cell vacuoles. During the cell elongation and subsequent maturation of leaves and fruit, there is little further biosynthesis. As the fruit reaches maturity, flavonoid concentrations in the fruit, including naringin, are lowered due to dilution effects (Hagen et al. 1966; Jourdan et al. 1985b; Vandercook and Tisserat 1989; Berhow and Vandercook 1991; Shaw et al. 1991; Castillo et al. 1992). Thus juice from early season grapefruit is generally more bitter than juice from late season grapefruit (Attaway 1977). Levels of naringin are much higher in the peel than in the juice sacs, and, as naringin is very soluble in water, the naringin content of commercially processed juice is dependent on the pressure on the fruit during the extraction process. Interestingly, the metabolic fate of naringin and the other flavanone glycosides in the plants is not known, and there is as yet no clue as to as to what functional role these compounds play in the plant themselves. Flavonoid composition in citrus has been extensively examined. The most comprehensive reviews on the composition of the citrus flavonoids are by Horowitz and Gentili (1977) and Kefford and Chandler (1970).

3.1 Analysis of Flavonoid Bitter Principles

The simplest and most direct way to assay for bitter principles in citrus and citrus products is by tasting. Unfortunately, there is a wide range of sensitivity to flavanone neohesperidoside bitterness (Guadagni et al. 1973). Some individuals are acutely sensitive to naringin, while others cannot detect it at even relatively high levels. This may be due to both genetic factors and/or variations in stereochemical composition (Gaffield et al. 1975). Due to this variation, taste panels are not a reliable measure of bitterness and other methods have been developed. Several of these methods have been reviewed previously. An excellent detailed review of citrus bitterness anlysis methodology was prepared by R.L Johnson in the book entitled *Analysis of Nonalcoholic Beverages* (1988). Other method reviews include Russell L. Rouseff's book *Bitterness in Foods and Beverages* (1990), and briefly in S.V. Ting and R.L. Rouseff's *Citrus Fruits and Their*

Products (1986) and Dan A. Kimball's *Citrus Processing: Quality Control and Technology* (1991).

3.1.1 Sample Preparation

There are a wide variety of methods to prepare samples for analysis by the methods discussed below. Many are needlessly complex. In some cases, juice samples can be analyzed directly after removing the particulates through centrifugation. Other tissue samples, including juice, can be extracted or mixed with organic solvents such as ethyl acetate, the methylchlorides, methanol, propanol etc., usually heated to increase the efficiency of extraction. As some of the flavonoid glycosides, such as hesperidin, are not very soluble in many solvents, dimethylsulphide or dimethylformamide can be sued in the extraction medium. One of the most effective methods is to extract tissues with a 1:1 mixture of dimethylsulphoxide and methanol (Berhow and Vandercook 1989). The DMSO can be removed by diluting the extract about tenfold with water and syringing the mixture through a small C-18 solid phase preparatory column. The flavonoids are retained on the column and, after rinsing with water, can be eluted with an appropriate solvent such as methanol. The samples can then be concentrated for spotting on TLC plates or filtered for injection on HPLC columns.

3.1.2 Colorimetric and Spectrophotometric Methods

The most convenient method to estimate the concentration of certain flavanones in citrus juices is the Davis Method (Davis 1947). It depends on the development of a yellow colour under basic conditions of the Davis reagent in which the flavanones are converted to the chalcone form. The relative absorbance at 427 nm of samples treated in this manner has been used as a measure of naringin content. However, this method is not selective and is rather a measure of the total flavanone glycosides in the sample (Horowitz and Gentili 1959; Fisher and Wheaton 1976). The Davis reagent can be used more specifically to estimate the concentration of individual flavanones colorimetrically after they have been separated by TLC, scraped off the plate and added to tubes containing the reagent (Fisher et al. 1966). The Davis Method can be used as a measure of the relative decrease in flavonoids and therefore in bitter flavonoids during a debittering process.

There are several other colorimetric methods in the literature including measuring the pink colour flavonoids developed in the presence of magnesium and hydrochloric acid (Kwierty and Braverman 1959), the red colour developed by flavonoids when treated with 1-nitroso-2-naphthol and nitric acid (Gengross and Renda 1966) and using an enzymatic colorimetric measurment with 2,2′-azino-bis-(3-ethylbenzthiazoline-6-sulphonic acid) (Arnao et al. 1990).

UV absorption can also be used to rapidly estimate flavonoid conent in samples comparable to the result obtained by the Davis method (Hendrickson and Edwards 1958). Sample is diluted/extracted with 2-propanol and filtered to obtain an extract that can be measured in a spectrophotometer. The difference between the absorbances of the extract at 295 and 290 nm is proportional to the flavonoid content.

3.1.3 Thin-Layer Chromatography

There are many published TLC protocols that can be used to separate flavonoids that have been extracted from tissue samples and juice. Methods include the use of silica, reverse phase or polyamide plates in a wide variety of solvent systems. Many of these solvent systems are published in a chapter on flavonoids in Kirchner's book on TLC (1967). The flavonoid spots on the developed plates can be visualized in a variety of maners, the most common method being the direct visualization under UV light using plates that have an acid-stable fluorescent indicatior added for visualization at 254 nm. The visualized spots can then be scraped and quantified by UV absorption. Another standard visualization method utilizes a spray of 2% sodium borohydride freshly prepared in methanol followed by fumigation with hydrogen chloride gas (Horowitz 1957).

Published TLC methods usually vary from one another mainly in the extraction procedures. We have found that one of the best solvent systems for developing silica plates is ethyl acetate-methylethylketone-formic acid-water (5:3:1:1) and that for developing polyamide plates is nitromethane-methanol (2:1), though there are many others. Several protocols have been published including Hagen et al. (1965), Mizelle et al. (1965), Fisher et al. (1966), Coffin (1971), Dass and Weaver (1972), Tatum and Berry (1973b), Albach and Wutscher (1988). Several grops have used TLC to examine flavonoid composition in *Citrus* species including: Albach et al. (1969), Albach and Redman (1969), Nishura et al. (1969, 1971a,b), Dass (1977), Dass et al. (1978), Kamiya et al. (1979) and Anis and Aminuddin (1981).

3.1.4 High-Performance Liquid Chromatography

HPLC is the method of choice to accurately determine both the composition and the absolute concentrations of the flavonoids in a given sample. Since HPLC was first described for use in flavonoid determination by Fisher and Wheaton (1976), its use has been expanded and improved to evalute flavonoids in extracts from fruit tissues, leaves, stems, roots, floral parts and juice. Extraction procedures are similar to those used in preparing samples for TLC, except that non-evaporating solvents such as dimethylsulphoxide and dimethylformamide can be utilized, which are excellent solvents for most of the flavonoids. Sample preparation can be a complex series of steps that includes some solid phase preparative separations or as simple as extraction, filtration and injection.

Most of the HPLC analyses are run using C-18 ODS reverse phase columns with either isocratic or gradient elutions. Isocratic systems include simple acetonitrile-water systems (80:20) (Fisher 1977) which is now usually enhanced by the addition of a small amount of 0.5% glacial acetic acid (Rouseff 1988). Del Rio et al. (1992) used an isocratic water-methanol-acetonitrile-acetic acid (15:2:2:1) or a methanol-0.01 M phosphoric acid isocratic system and Mouly et al. (1993) have used a water-acetonitrile-tetrahydrofuran-acetic acid (80:16:3:1) mixture to develop the columns. Ternary gradients were used by Ting et al. (1979), Park et al. (1983) and Sendra et al. (1988). Binary gradients of methanol and 0.01 M phosphoric acid have been used by Berhow and Vandercook (1989) and del Rio et al. (1992). Most of these methods successfully resolve the flavonoids, even the neohesperidoside and rutinoside isomers. A binary gradient of [water-methanol-acetic acid (90:10:0.5)] and methanol on a β-cyclodextrin column was used to separate diastereomeric isomers of naringin by Krause and Galensa (1991). Identification and quantification are usually achieved by absorbance at 280–285 nm and comparison to runs of known standards, but retention times of flavonoids in complex mixtue can vary considerably from those of separately run standards. The resolution limit is on the order of 1 microgram, but samples can be concentrated to overcome that limit. The new technology of Diode Array detectors (Berhow and Vandercook 1989) and Mass Spectrometers (Hadj-Mahammed and Meklati 1987) couplted to HPLC systems yield more accurate peak assignments when analyzing samples.

3.1.5 Radioimmunoassay (RIA) and Enzyme-Linked Immunoassay (ELISA or EIA)

Jourdan et al. (1982) reported the development of a RIA to detect the presence of the bitter flavanone neohesperdosides. The assay uses an immobilized antibody raised in rabbit against a naringin-bovine serum albumin conjugate which is specific for the flavonoid neohesperidosides. Samples are added to wells containing the immobilized enzyme bound to the antibody. The amount of bound material is determined by saturating the remaining antibodies with either ^{3}H-labelled or ^{125}I-labelled naringin. This method can be used to detect naringin down to 2 nanograms. The assay combines both high sensitivity, and rapid, easily performed assays. RIAs have been used to analyze naringin levels in *Citrus* tissue (Jourdan et al. 1985a,b).

A non-radioactive form of this assay, the enzyme-linked immunoassay which utilizes naringin linked to alkaline phosphatase was developed by Jordan et al. (1984) and used by Matsumoto and Okudai (1991a,b). While immunoassays are extremely sensitive and fast, they do have a number of interfering agents and cannot accurately discriminate between the four different flavanone neohesperidosides. In addition, the components for the assay – the anti-naringin antibodies, the radioactively labelled naringin or the alkaline phosphatase-linked naringin, are not commonly available.

3.1.6 Capillary Electrophoresis

A relatively new technique, capillary electrophoresis (CE) is becoming increasingly recognized as an important analytical separation technique. It has the advantage of being fairly rapid (runs of about 10 min), simple, and similar to the HPLC methodology. One of the first protocols was published for the separation of flavonol-3-glycosides from *Ginkgo biloba* (Pietta et al. 1991). Seitz et al. (1992) published a CE protocol for the separation of flavonoids from *Sambuci flos*. As sample preparation and chromatography conditions are fairly standardized, these techniques could easily be modified for the analysis of citrus flavonoids.

3.1.7 Gas Chromatography-Mass Spectroscopy

Because the flavonoid glycosides are non-volatile, GC has not routinely been used for flavonoid analysis. However, Creaser et al. (1992) have developed a method to analyze the hydrolyzed flavanone content by GC-MS. The method involves solvent extraction of the tissues, hydroylsis of the flavonoids to their respective aglycones, trimethylsilation, and analysis by GC-MS. While the method will accurately identify and quantify the flavonoid aglycones present in the sample, it cannot determine the relative amounts of the neohesperidosides and rutinosides, and would therefore only be able to give an estimation of the amount of bitter flavonoids in a sample.

3.2 Sources of Flavonoid Standards

Over 50 different flavonoids have been isolated and identified from the tissues of citrus and its related species. Many of these are not commercially available. However, all the large chemical companies, such as Sigma, Aldrich, Fluka etc., sell a few of the more common flavonoids such as naringin, hesperidin, narigenin, hesperetin, rutin, quercetin and a few others. There are at least two chemical companies that sell a much wider selection of flavonoid chromatographic standards, including all four of the bitter flavanone neohesperidosides and all four of the flavanone rutinosides. They are: Extrasynthese SA (BP 62, 21 Lyon Nord 69730 Gernay, France) and Carl Roth GmbH + CoKG Chemische Fabrik (Schoemperlenstrasse 1–5, Postfach 21 11 62, D 76185 Karlsruhe, Germany, distributed in North America by Atomergic Chemetals Corp., 91 Carolyn Blvd., Farmingdale, NY 11735, USA).

References

Albach RF, Redman GH (1969) Composition and inheritance of flavanones in citrus fruit. Phytochemistry 8:127–143
Albach RF, Wutscher HK (1988) Flavanone content of whole grapefruit and juice as influenced by fruit development. J Rio Grande Val Hortic Soc 41:89–95
Albach RF, Juarez AT, Lime BJ (1969) Time of naringin production in grapefuit. J Am Soc Hortic Sci 94:605–609

Alford AR, Bentley MD (1986) Citrus limonoids as potential antifeedants for the spruce budworm (Lepidoptera: Torticidae). J Econ Entomol 79:35–38
Anis M, Aminuddin (1981) Flavonoid patterns of leaves of some *Citrus* species and their hybrids. Plant Biochem J 8:56–60
Arnao MB, Casas JL, del Rio JA, Acosta M, Garcia-Canovas F (1990) An enzymatic colorimetric method for measuring naringin using 2,2′-azino-bis-(3-ethylbenzthiazoline-6-sulfonic acid) (ABTS) in the presence of peroxidase. Anal Biochem 185:335–338
Attaway JA (1977) Factors influencing the flavor of grapefruit juice. Proc Int Soc Citricult 3:816–820
Berhow MA, Vandercook CE (1989) Biosynthesis of naringin and prunin in detached grapefruit. Phytochemistry 28:1627–1630
Berthow MA, Vandercook CE (1991) Sites of naringin biosynthesis in grapefruit seedlings. J Plant Physiol 138:176–179
Castillo J, Benavente O, del Rió JA (1992) Naringin and neohesperidin levels during development of leaves, flower buds, and fruits of *Citrus aurantium*. Plant Physiol 99:67–73
Coffin DE (1971) A method for the isolation and identification of the flavanone glycosides of citrus juices. J Agric Food Chem 19:513–516
Creaser CS, Koupai-Abyazani MR, Stephenson GR (1992) Gas chromatographic-mass spectrometric characterization of flavanones in citrus and grape juices. Analyst 117:1105–1109
Dass HC (1977) Phylogenetic studies in *Citrus reticulata* and related types. Euphytica 26:805–810
Dass HC, Weaver GM (1972) Cellulose thin-layer chromatography of phenolic substances. J Chromatogr 67:105–111
Dass HC, Randhawa GS, Prakash D (1978) Leaf flavonoid patterns of some importan *Citrus* species. Indian J Exp Bot 16:62–65
Davis WB (1947) Determination of flavanones in citrus fruits. Anal Chem 19:467–478
del Rio JA, Benavente O, Castillo J, Borrego F (1992) Neodiosmin, a flavone glycoside of *Citrus aurantium*. Phytochemistry 31:723–724
Fisher JF (1968) A procedure for obtaining radioactive naringin from grapefruit leaves fed L-phenylalanine-^{14}C. Phytochemistry 7:769–771
Fisher JF (1973) Fluorometric determination of limonin in grapefruit and orange juice. J Agric Food Chem 21:1109–1110
Fisher JF (1977) Review of quantitative analyses for limonin, naringin, naringenin rutinoside, hesperidin, and neohesperidin dihydrochalcone in citrus juice by high performance liquid chromatography. Proc Int Soc Citricult 3:813–816
Fisher JF, Wheaton TA (1976) A high-pressure liquid chromatographic method for the resolution and quantitation of naringin and naringenin rutinoside in grapefruit juice. J Agric Food Chem 24:898–899
Fisher JF, Nordby HE, Kew TJ (1966) A thin-layer chromatographic-colorimetric method for determining naringin in grapefruit. J Food Sci 31:947–950
Fong CH, Hasegawa S, Herman Z, Ou P (1991) Biosynthesis of limonoid glucosides in lemon (*Citrus limon*). J Sci Food Agric 54:393–398
Fong CH, Hasegawa S, Miyake M, Ozaki Y, Coggins CW, Atkin DR (1992) Contents of limonoids and limonin 17-β-D-glucopyranoside in fruit tissue of Valencia orange during fruit growth and maturation. J Agric Food Chem 40:1178–1181
Gaffield W, Lundin RE, Gentili B, Horowitz RM (1975) C-2 sterochemistry of naringin and its relation to taste and biosynthesis in maturing grapefruit. Bioorg Chem 4:259–269
Gengross O, Renda N (1966) Occurance and quantitative estimation of naringin in citrus species. Justus Liebigs Ann Chem 691:186–189
Guadagni DG, Maier VP, Turnbaugh JG (1973) Effect of some citrus juice constituents on taste thresholds for limonin and naringin bitterness. J Sci Food Agric 24:1277–1288
Guadagni DG, Maier VP, Turnbaugh JG (1974a) Some factors affecting sensory thresholds and relative bitterness of limonin and naringin. J Sci Food Agric 25:1199–1205
Guadagni DG, Maier VP, Turnbaugh JG (1974b) Effect of subthreshold concentrations of limonin, naringin and sweeteners on bitterness perception. J Sci Food Agric 25: 1349–1354

Hadj-Mahammed M, Meklati BY (1987) Qualitative determination of polymethoxylated flavones in Valencia orange peel oil and juice by LC-UV/VIS and LC-MS techniques. Lebensm-Wiss Technol 20:111-114

Hagen RE, Dunlap WJ, Mizelle JW, Wender SH, Lime BJ, Albach RF, Griffiths FP (1965) A chromatographic-fluorometric method for detemination of naringin, naringenin rutinoside, and related flavanone glycosides in grapefruit juice and juice sacs. Anal Biochem 12:472-482

Hagen RE, Dunlap WJ, Wender SH (1966) Seasonal variation of naringin and certain other flavanone glycosides in juice sacs of Texas Ruby Red grapefruit. J Food Sci 31:542-547

Hahlbrock K, Scheel D (1989) Physiology and molecular biology of phenylpropanoid metabolism. Annu Rev Plant Physiol Plant Mol Biol 40:347-369

Hasegawa S, Herman Z (1992) Biosynthesis of limonoids in *Citrus*. In: Petroski RJ, McCormick SP (eds) Secondary-metabolite biosynthesis and metabolism. Plenum Press, New York, pp 305-317

Hasegawa S, Maier VP (1981) Some aspects of citrus biochemistry and juice quality. Proc Int Soc Citricult 2:914-918

Hasegawa S, Pelton VA (1983) Reduction of nomilin bitterness in citrus juices and juice serum with *Arthrobacter globiformis* cells immobilized in acrylamide gel. J Agric Food Chem 31:178-180

Hasegawa S, Bennett RD, Verdon CP (1980) Limonoids in citrus seeds: origin and relative concentration. J Agric Food Chem 28:922-925

Hasegawa S, Bennett RD, Maier VP (1984) Biosynthesis of limonoids in *Citrus* seedlings. Phytochemistry 23:1601-1603

Hasegawa S, Herman Z, Orme E, Ou P (1986) Biosynthesis of limonoids in *Citrus*: sites and translocation. Phytochemistry 25:2783-2785

Hasegawa S, Bennett RD, Herman Z, Fong CH, Ou P (1989) Limonoid glucosides in citrus. Phytochemistry 28:1717-1720

Hasegawa S, Ou P, Fong CH, Herman Z, Coggins CW, Atkin DR (1991) Changes in the limonoate A-ring lactone and limonin 17-β-D-glucopyranoside content of navel oranges during fruit growth and maturation. J Agric Food Chem 39:262-265

Hashinaga F, Ejima H, Nagahama H, Itoo S (1977) Studies on the limonoids in citrus fruits. I. Seasonal changes of limonoid components in ponkan, tankan, early satsuma mandarin and natsudaidai fruits. Kagoshima Daigaku Nogakubu Gakujutsu Hokoku 27:171-180

Heller W, Forkmann G (1988) Biosynthesis. In: Harborne JB (ed) The flavonoids. Chapman and Hall, London, pp 399-425

Hendrickson R, Edwards GJ (1958) Ultraviolet absorption technique to determine the naringin content of grapefruit juice. Proc Fla State Hortic Soc 71:194-198

Herman Z, Hasegawa S, Fong CH, Ou P (1989) Limonoids in *Citrus ichangensis*. J Agric Food Chem 37:850-851

Herman Z, Fong CH, Hasegawa S (1992) Analysis of Limonoids in Citrus Seeds. In: Linskens HF, Jackson JF (eds) Seed analysis. Modern methods of plant analysis new series, vol 14. Springer, Berlin Heidelberg, New York, pp 361-375

Hidaka T, Omura M, Ugaki M, Tomiyama M, Kato A, Ohshima M, Motoyoshi F (1990) *Agrobacterium*-mediated transformation and regeneration of *Citrus* spp. from suspension cells. Jpn J Breed 40:199-207

Horowitz RM (1957) Detection of flavanones by reduction with sodium borohydride. J Org Chem 22:1733-1734

Horowitz RM, Gentili B (1959) Use of the Davis method to estimate flavanones. Food Res 24:757-759

Horowitz RM, Gentili B (1977) Flavonoid constituents of *Citrus*. In: Nagy S, Shaw PE, Velduis MK (eds) *Citrus* science and technology. AVI Publishing, Westport, CT, pp 397-426

Horowitz RM, Gentili B (1986) Dihydrochalcone sweeteners from citrus flavanones. In: O'Brien-Nabors L, Gelardi RC (eds) Alternative sweeteners. Marcel Dekker, New York, pp 135-153

Johnson RL (1988) Bitterness in fruit juices: assay of bitter principles in citrus juices. In: Linskens HF, Jackson JF (eds) Analysis of nonalcoholic beverages. Springer, Berlin Heidelberg New York, pp 80-103

Jourdan PS, Mansell RL, Weiler EW (1982) Radioimmunoassay for the citrus bitter principle, naringin, and related flavonoid-7-O-β-neohesperidosides. J Med Plant Res 44:82–86

Jorudan PS, Mansell RL, Oliver DG, Weiler EW (1984) Competitive solid phase enzyme-linked immunoassay for the quantification of limonin in citrus. Anal Biochem 138:19–24

Jourdan PS, Weiler EW, Mansell RL (1985a) Naringin levels in citrus tissues. I. Comparison of different antibodies and tracers for the radioimmunoassay of naringin. Plant Physiol 77:896–902

Jourdan PS, Mclntosh CA, Mansell RL (1985b) Naringin levels in citrus tissues. II. Quantitative distribution of naringin in *Citrus paradisi* Macfad. Plant Physiol 77:903–908

Kamiya S, Esaki S, Konishi F (1979) Flavonoids in citrus hybrids. Agric Biol Chem 43:1529–1536

Kefford JF, Chandler BV (1970) The chemical constituents of *Citrus* fruits. Academic Press, New York

Kimball DA (1991) *Citrus* processing: quality control and technology. Van Nostrand Reinhold, New York

Kirchner JR (1967) Thin layer chromatography. John Wiley, New York

Krause M, Galensa R (1991) High-performance liquid chromatography of diastereomeric flavanone glycosides in *Citrus* on a beta-cyclodextrin-bonded stationary phase (Cyclobond I). J Chromatogr 588:41–45

Kruger AJ, Colter CE (1972) Gas chromatographic identification of limonin in citrus juice. Proc Fla State Hortic Soc 85:206–210

Kwierty A, Braverman JB (1959) Critical evaluation of the cyanidin reaction for flavonoid compounds. Bull Res Counc Isr Sect C 7:187–196

Lam LKT, Hasegawa S (1989a) Inhibition of benzo[a]pyrene-induced forestomach neoplasia in mice by citrus limonoids. Nutr Cancer 12:43–47

Lam LKT, Li Y, Hasegawa S (1989b) Effects of citrus limonoids on glutathione S-transferase activity in mice. J Agric Food Chem 37:878–880

Liu Y, Alford AR, Rajab MS, Bentley MD (1990) Effects and modes of action of citrus limonoids against *Leptinotarsa decemlineata*. Physiol Entomol 15:37–45

Maier VP, Beverly GD (1968) Limonin monolactone, the nonbitter precursor responsible for delayed bitterness in certain citrus juices. J Food Sci 33:488–492

Maier VP, Grant ER (1970) Specific thin-layer chromatography assay of limonin, a citrus bitter principle. J Agric Food Chem 18:250–252

Maier VP, Hasegawa S, Hera E (1969) Limonin D-ring-lactone hydrolase. A new enzyme from *Citrus* seeds. Phytochemistry 8:405–407

Mansell RL, Weiler EW (1980) Radioimmunoassay for the determination of limonin in citrus. Phytochemistry 19:1403–1407

Matsumoto R, Okudai N (1991a) Early evaluation of citrus bitter component, flavanone neohesperidosides by enzyme immunoassay using anti-naringin antibody. J Jpn Soc Hortic Sci 60:191–200

Matsumoto R, Okudai N (1991b) Inheritance of flavanone neohesperidoside in *Citrus*. J Jpn Soc Hortic Sci 60:201–207

Miller EG, Fanous R, Rivera-Hidalgo F, Binnie WH, Hasegawa S, Lam LKT (1989) The effect of citrus limonoids on hamster buccal pouch carcinogenesis. Carcinogenesis 10:1535–1537

Miller EG, Gonzales-Sanders AP, Courillon AM, Wright JM, Hasegawa S, Lam LKT (1992) Inhibition of hamster buccal pouch carcinogenesis by limonin 17-β-D-glucopyranoside. Nutr Cancer 17:1–7

Miyake M, Ayano S, Ozaki Y, Maeda H, Ifuku Y, Hasegawa S (1991) Extraction of neutral limonoids from citrus seeds. Nippon Nogeikagaku Kaishi 65:987–992

Mizelle JW, Dunlap WJ, Hagen RE, Wender SH, Lime BJ, Albach RF, Griffiths FP (1965) Isolation and identification of some flavanone rutinosides of the grapefruit. Anal Biochem 12:316–324

Mouly P, Gaydou EM, Estrienne J (1993) Column liquid chromatographic determination of flavanone glycosides in *Citrus*. J Chromatogr 634:129–134

Nishiura M, Esaki S, Kamiya S (1969) Flavonoids in *Citrus* and related genera. I. Distribution of flavonoid glycosides in *Citrus* and *Poncirus*. Agric Biol Chem 33:1109–1118

Nishiura M, Kamiya S, Esaki S, Ito F (1971a) Flavonoids in *Citrus* and related genera. II. Isolation and identification of isonaringin and neoeriocitrin from *Citrus*. Agric Biol Chem 35:1683–1690
Nishiura M, Kamiya S, Esaki S (1971b) Flavonoids in *Citrus* and related genera. III. flavonoid pattern and *Citrus* taxonomy. Agric Biol Chem 35:1691–1706
Ou P, Hasegawa S, Herman Z, Fong CH (1988) Limonoid biosynthesis in the stem of *Citrus limon*. Phytochemistry 27:115–118
Park GL, Avery SM, Byers JL, Nelson DB (1983) Identification of bioflavonoids from *Citrus*. Food Technol 37:98–105
Pietta PG, Mauri PL, Rava A, Sabbatini G (1991) Application of micellar electrokinetic capillary chromatography to the determination of flavonoid drugs. J Chromatogr 549:367–373
Rouseff RL (1982) Nomilin, a new bitter component in grapefruit juice. J Agric Food Chem 30:504–507
Rouseff RL (1988) Differentiating citrus juices using flavanone glycoside concentration profiles. In: Nagy S, Attaway JA, Rhodes ME (eds) Adulteration of fruit juice beverages (Food science and technology, vol 30) Marcel Dekker, New York, pp 49–66
Rouseff RL (ed) (1990) Bitterness in foods and beverages. (Developments in Food Science, vol 18) Elsevier, New York
Rouseff RL, Fisher JF (1980) Determination of limonin and related limonoids in citrus juices by high performance liquid chromatography. Anal Chem 52:1228–1233
Rouseff RL, Mansell RL (1982) Comparison of high performance liquid chromatographic and radioimmunoassay limonin values from commercial grapefruit juice. Proc Fla State Hortic Soc 95:249–252
Rouseff RL, Matthews RF (1984) Nomilin, taste threshold and relative bitterness. J Food Sci 49:777–779
Rouseff RL, Nagy S (1982) Distribution of limonoids in citrus seeds. Phytochemistry 21:85–90
Seitz U, Oefner PJ, Nathakarnkitkool S, Popp M, Bonn GK (1992) Capillary electrophoretic analysis of flavonoids. Electrophoresis 13:35–38
Sendra JM, Navarro JL, Izquierdo L (1988) C18 solid-phase and high performance liquid chromatography/ultraviolet diode array determination of fully methoxylated flavones in citrus juices. J Chromatogr Sci 26:443–448
Shaw PE (1986) Use of a simple solvent optimization program to improve separation of limonin in citrus juices. J Chromatogr Sci 24:364–366
Shaw PE (1990) Cyclodextran polymers in removal of bitter compounds in citrus juices. In: Rouseff R (ed) Bitterness in foods and beverages. Elsevier, New York, pp 309–324
Shaw PE, Wilson CW (1984) A rapid method for determination of limonin in citrus juices by high performance liquid chromatography. J Food Sci 49:1216–1218
Shaw PE, Calkins CO, McDonald RE, Greany PD, Webb JC, Nisperos-Carriedo MO, Barros SM (1991) Changes in limonin and naringin levels in grapefruit albedo with maturity and the effects of gibberellic acid on these changes. Phytochemistry 30:3215–3219
Stafford HA (1990) Flavonoid metabolism. CRC Press, Boca Raton, FL
Tatum JH, Berry RE (1973a) Method for estimating limonin content of citrus juices. J Food Sci 38:1244–1246
Tatum JH, Berry RE (1973b) Method for determining naringin content in grapefruit juice. J Food Sci 38:340–341
Ting SV, Rouseff RL (1986) *Citrus* fruits and their products. (Food Science and Technology, vol 18) Marcel Dekker, New York, pp 135–153
Ting SV, Rouseff RL, Dougherty MH, Attaway JA (1979) Determination of some methoxylated flavones in citrus juices by high performance liquid chromatography. J Food Sci 44:69–71
Van Beek TA, Blaakmeer A (1989) Determination of limonin in grapefruit juice and other citrus juices by high-performance liquid chromatography. J Chromatogr 464:375–386
Vandercook CE, Tisserat B (1989) Flavonoid changes in developing lemons grown in vivo and in vitro. Phytochemistry 28:799–803
Vardi A, Bleichman S, Aviv D (1990) Genetic transformation of *Citrus* protoplasts and regeneration of transgenic plants. Plant Sci 69:199–206
Weiler EW, Mansell RL (1980) Radioimmunoassay of limonin using a tritiated tracer. J Agric Food Chem 28:543–545

Widmer WW (1991) Improvements in the quantitation of limonin in citrus juice by reversed-phase high-performance liquid chromatography. J Agric Food Chem 39:1472–1476
Widmer WW, Rouseff RL (1991) Quantitative analysis of limonin in grapefruit juice using an enzyme-linked immunoassay: collaborative study. J Assoc Off Anal Chem 74:513
Wilson KW, Crutchfield CA (1968) Spectrophotometric determination of limonin in orange juice. J Agric Food Chem 16:118–124

Phenolic Content and L-Phenylalanine Ammonia-Lyase Activity in Peach Fruit

N. Kubota

1 Introduction

There are a wide variety of secondary metabolites of phenolic compounds found throughout the plant kingdom. They are important not only physiologically but also pathologically. Some phenolic compounds such as tannins have pharmacological and emollient properties and are used therapeutically or for the tanning of leather. Numerous studies of the quantitative and qualitative variations in phenolic compounds from various higher plants have been undertaken.

In general, phenolic compounds in fruit contribute to their sensory qualities such as colour, aroma, bitterness and astringency (Ozawa et al. 1987; Macheix et al. 1990). Judging from their ployphenol content (Joslyn and Goldstein 1964; Van Buren 1970), peach fruit (*Prunus persica* Batsch.) is usually less astringent than other fruits. However, the degree of astringency can increase due to factors such as meteorological and growing conditions, in which case the commercial value of the peach fruit decreases (Sistrunk 1985; Senter and Callahan 1990). L-Phenylalanine ammonia-lyase (PAL;EC 4.3.1.5), the enzyme that catalyzes the conversion of L-phenylalanine to *trans*-cinnamic acid, was demonstrated first in barley by Koukol and Conn (1961). Since then, numerous studies of PAL in various plants have been reported. The enzyme is widely distributed in higher plants, including their fruit, and is important in the biosynthesis of phenolic compounds (Camm and Towers 1973). The concentrations of such compounds and the PAL activity in peach fruit have been investigated by many researchers (Guadagni and Nimmo 1953; Reeve 1959; Craft 1961; Goldstein and Swain 1963; Aoki et al. 1971), but information about the relationship between concentration or PAL activity and astringency is scanty.

In this chapter, methods of quantitative and qualitative analysis of phenolic compounds in peach fruit, and of measurement of PAL activity will be described with special reference to astringency. The relationship between the phenolic content and PAL activity in peach fruit, and some factors affecting the content and activity, will be discussed.

Modern Methods of Plant Analysis, Vol. 18
Fruit Analysis
Edited by H.-F. Linskens and J.F. Jackson

2 Phenolic Content and Phenolic Composition of Peach Fruit

2.1 Analysis of Phenolic Compounds

2.1.1 Total and High-Molecular-Weight Phenolics

According to Joslyn and Goldstein (1964), the Folin-Denis method, published in 1912, was first intended for the measurement of tannins in wine and whiskey, and then used for the assay of tannins in fruits (Swain and Hillis 1959; Craft 1961). Various methods for phenolic compound analysis have been reported (Joslyn and Goldstein 1964; Hartley 1987; Karchesy 1989; Hagerman 1989), but the Folin-Denis method is still very useful for the measurement of phenolic compounds in fruits.

Techniques for the analysis of total phenolic compounds in peach fruit based on the Folin-Denis method (Swain and Hillis 1959) are described below. All chemicals are reagent-grade commercial products. Phenolic compounds in peach fruit are extracted with distilled water, because the soluble compounds are thought to give rise to the sensation of astringency. Ten g of mesocarp tissue (edible portion) of peach fruit is boiled in 30 ml of distilled water for 20 min and then blended with the water in a mortar. The homogenate is filtered by suction through a Büchner funnel. After dilution of the filtrate with distilled water to an appropriate concentration, the total phenolic contents are measured by the Folin-Denis method. To 5 ml of the extract, 5 ml of Folin reagent is added, and the mixture is shaken. After 3 min, 5 ml of 10% Na_2CO_3 (anhydrous) is added, and the mixture is shaken again and left for 1 h. After centrifugation or filtration of the mixture, its absorbance at 700 nm in a 1-cm glass cell is measured. The Folin reagent is prepared as follows. A mixture of 25 g $Na_2WO_4 \cdot 2H_2O$, 5 g $P_2O_5 \cdot 24MoO_3 \cdot nH_2O$, 12.5 ml H_3PO_4, and 188 ml of distilled water is heated to dissolution for 2 h under reflux conditions, and after the mixture cools, its volume is brought to 1000 ml with distilled water. The content of total phenolic compounds in fruits is generally expressed as milligrams of D-catechin per gram or 100 g fresh weight (FW). The standard curve of D-catechin is linear to 20 μg per 5 ml of distilled water. In other kinds of experiments, 80% (v/v) methanol or ethanol solutions is used for extraction of the phenolic compounds. In that case, the concentration of methanol or ethanol in the extracts must be lowered to less than 5% by evaporation or dilution before analysis, because high concentration of these substances would affect the measurement of absorbance (Swain and Hillis 1959).

After precipitation of tannins in the extracts obtained as described above by the addition of gelatin, the amount of phenolic compounds of high-molecular-weight fraction (HMWF) in a sample can be measured by the Folin-Denis method. The procedure is as follows. For removal of tannins from the extract, 5 ml of gelatin solution (2.5 g of a sheet of gelatin is dissolved in water with heating and the solution is brought to 100 ml with distilled water) is added to 10 ml of the extract. Then 10 ml of an acid-saturated salt solution (saturated solution of NaCl : H_2SO_4 =

975:25, v/v) is added, followed by 1 g of kaolin. The mixture is stirred vigorously with a glass rod and left until a precipitate forms. The supernatant is filtered with paper, and the absorbance of the filtrate is measured by the Folin-Denis method as described above. The value obtained is subtracted from the total phenolic content and designated as the HMWF of the phenolic compounds.

2.1.2 Condensed Tannins

The ability of tannins to precipitate water-soluble proteins is the main activity by which various naturally occurring polyphenols are defined as tannins. Several methods for measurement of tannins based on these activities have been developed (Bate-Smith 1973; Porter and Woodruffe 1984; Hagerman 1989). Methods for the assay of relative astringency (RA) and relative affinity to methylene blue (RMB) have been used frequently in pharmacological studies for the measurement of the tannin content of plant extracts (Okuda et al. 1985).

The RA method is used to assay condensed tannins or proanthocyanidin polymers in plant tissues. To the supernatant, a tannin is added, causing precipitation of tannin-protein complexes, and the precipitate is removed by centrifugation. The intensity of colour of the haemoglobin remaining is compared with that of a standard solution of haemoglobin and tannic acid (Porter and Woodruffe 1984). The approach used in my group is essentially that of Porter and Woodruffe (1984). A standard curve is obtained as follows: exactly 0, 0.1, 0.2, 0.3 or 0.4 ml of a solution of 1000 ppm of tannic acid is put into five centrifuge tubes. Distilled water is added to each tube to bring the volume to 0.4 ml, followed by 0.4 ml of diluted blood and finally 0.2 ml of phosphate buffer (0.2 M KH_2PO_4:0.2 M $Na_2HPO_4 \cdot 12H_2O$ = 87.7:12.3, v/v. pH 6.0). Before analysis, fresh human blood is haemolyzed by dilution with distilled water to make a haemoglobin solution with an absorbance of about 2.5 at 578 nm. The mixture in each tube is vigorously shaken and left for 1 h. After centrifugation of the mixture at 12 000 *g* for 10 min, the absorbance of the supernatant at 578 nm is measured in a 0.5-cm glass cell. Extracts from mesocarp tissue of peach fruit also are analyzed in the same way. The RA is based on the amount of tannic acid needed for a 50% decrease in the absorbance by the test solution to which 0.4 ml distilled water was added. The following equation is used. RA = X (mg)/Y (mg), in which X is the tannic acid content at which the absorbance of the tannic acid solution had decreased to 50%, and Y is the weight of the sample the absorbance of which had decreased to 50% (see Fig. 1). Condensed tannins in peach fruit are extracted as follows: three separate batches of 5 g of mesocarp tissue of peach fruit each time is blended in a blender with 10 ml of 80% acetone. The homogenate is filtered through a Büchner funnel with suction. The three extracts are collected, evaporated to dryness and dissolved in 2.5 ml of distilled water. In all of these processes, reagent-grade chemicals are used.

Okuda et al. (1985) demonstrated that the RA and RMB values, based on those of geraniin, can be calculated reproducibly and rapidly from a small amount of

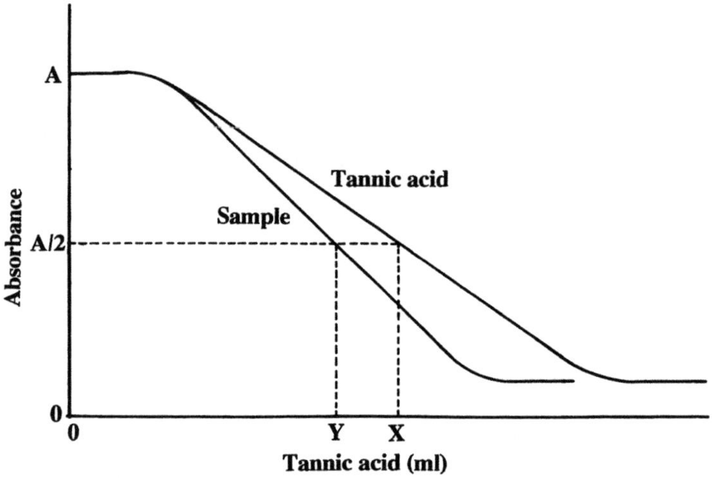

Fig. 1. Standard curve for assay of the content of condensed tannins by the RA method

sample for estimation of the tannin content of plant extracts. This method has been used for the evaluation of the basal activity of 84 tannins and related compounds. RMB based on geraniin is larger than RA based on geraniin for hydrolyzable tannins and smaller than RA based on geraniin for condensed tannins.

2.1.3 Phenolic Composition

Based on the results of paper chromatography, thin-layer chromatography (TLC) and high-pressure liquid chromatography (HPLC), several monomeric phenols (catechin, epicatechin, chlorogenic acid, isochlorogenic aicd, and neo-chlorogenic acid) have been found in mature and immature peach fruits (Nakabayashi 1968; Senter and Callahan 1990). Gallocatechin and epigallocatechin also are present in peach fruit (Macheix et al. 1990). Of these methods for chromatography, HPLC is perhaps the most useful for quantitative and qualitative assays of phenolic compounds; suitable procedures have been described in detail by Schwarzenbach (1982) and Hartley (1987).

The analysis by HPLC of phenolic compounds in peach fruit in our laboratory is done as follows. All chemicals are reagent-grade products. Three separate batches of 5 g each of mesocarp tissue are blended in a mortar with 10 ml of 80% (v/v) acetone, and the homogenate is filtered through a Büchner funnel with suction. The three extracts are collected, centrifuged to dryness at below 30 °C, lyophilized and stored until analysis. After dissolution of the extract, 1 ml of 50% methanol is added to each sample, and the solution is centrifuged for 10 min at 12 000 *g*. Portions (2 µl) of the supernatant are separated into components by HPLC. Authentic phenolic compounds are analyzed also. The unknown compounds are identified by comparison of spectra and retention times of peaks with the authentic compounds. The HPLC system is as follows: pump, Hitachi L-4000; detector, Hitachi L-6000, UV 280 nm; column, Hibar LicroCART 250–4; packing, Licrospher 100 RP-18 (5 μm), ODS type; column temp., 40 °C; mobile phase, phosphate buffer [(10 mM H_3PO_4:10 mM KH_2PO_4 = 1:1, v/v): methanol = 88:12, v/v], 1 ml/min; integrator, Hitachi D-2000.

2.2 Relationship Between Degree of Astringency Based on Organoleptic Tests and Phenolic Content

It is difficult to express exactly the degree of astringency, because unlike sweet and sour sensations, the sensation of astringency is not a true taste but is caused by astringent action. Perception of astringency and the threshold for perception differ from person to person. Joslyn and Goldstein (1964) reported that differences in the degree of astringency between peach fruit containing 69 and 111 mg of phenolic compounds were detected by 12 of 14 experts, but that differences between peaches containing 151 and 181 mg were detected by only 6 of 12 experts.

In our laboratory, we attempted to identify the relationship between the degree of astringency based on organoleptic tests and the content of total phenolic compounds in peach fruit (Kubota el al. 1986). The fruits were harvested at commercial maturity from Sanyo-suimitsu peach trees grown under eight different conditions, including soil moisture content and rootstock. The organoleptic tests for astringency were carried out by six experts selected from the staff of the Japan Food Research Laboratories Foundation, Tokyo, for their ability to distinguish among five tannic acid concentrations ranging from 0 to 60 mg per 100 ml of water. The scoring system for the evaluation of astringency was as follows. A score of 5 was given to one of eight groups of fruit in which the content of total phenolic compounds was closest to the mean value of the eight samples tested. Scores 1, 3, 7 and 9 were equivalent to very low, low, high and very high degrees of astringency, respectively. The mean value given by the six experts for each sample was calculated. The total phenolic content in each sample was assayed by the Folin-Denis method as described above. A large amount of total phenolic compounds resulted in a high degree of astringency, and the correlation between these properties was significant at the 5% level (Fig. 2).

As described in detail later, a large amount of total phenolics in mature peach fruit resulted in a large HMWF irrespective of the cultivar. It can be inferred that the size of the HMWF may be related more closely to astringency than the total phenolic contents, because astringency is due to the presence of phenolic compounds with a certain degree of polymerization, corresponding to a mean molecular weight of between 500 and 3000 (Van Buren 1970). Simple phenolic compounds, such as catechins, anthocyanins and hydroxycinnamic acids, do not cause astringency, unlike flavan dimers. Condensed tannins also are important in fruit astringency (Sistrunk 1985). However, the molecular weight of phenolics in the HMWF assayed as described above by the Folin-Denis method has not been

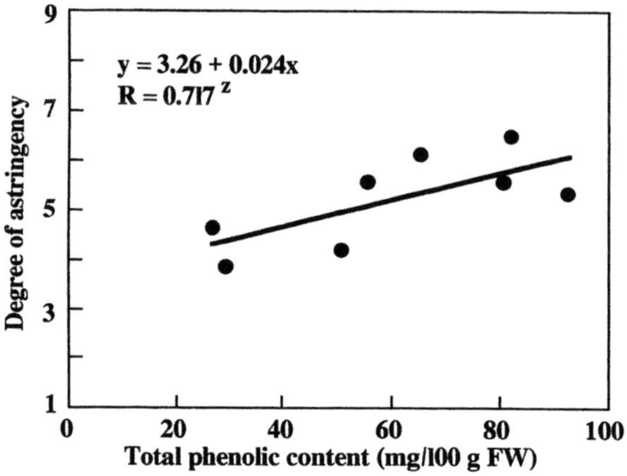

Fig. 2. Relationships between the content of total phenolic compounds and the degree of astringency based on organoleptic tests in peach fruit[z], $p = 0.05$

Table 1. Effects of soil moisture stress during stage 3 of fruit growth on fruit weight, total soluble solids and phenolic contents at harvest, and PAL activity of Sanyo-suimitsu peach fruit (Kubota and Kudo 1992)

Treatment[a] (pF)	Fruit weight (g)	Total soluble solids (°Brix)	Phenolic content		PAL activity[b]	
			mg/100 g FW	mg/fruit	*t*-Cinnamic acid (nmol/g FW·h)	Radioactivity incorporated into cinnamic acid from [^{14}C]-phenylalanine[c]
≦1.5	187.9a[d]	13.3b	55.1b	102.3b	N.D.[e]	5.1d
1.5–2.0	157.0b	13.9b	44.8b	70.7c	N.D.	10.7c
2.0–2.5	111.0d	16.4a	114.6a	127.7a	N.D.	45.6b
≧2.5	129.2c	16.0a	110.6a	143.9a	N.D.	64.5a

[a] Control of soil moisture was started on June 18.
[b] Measured 7 days after the start of treatment.
[c] $(\text{dpm} \times 10^3)/(\text{g Fw}\cdot\text{h})$
[d] Average values were analyzed with Duncan's multiple range test. Values within column followed by the same letter are not significantly different at $p = 0.01$.
[e] Not detected.

estimated. Relationships between the total phenolic content and the amounts of catechin and chlorogenic acids in peach fruit have been found, but the relationship between the total phenolic content and the RA depended on the cultivars. This result may mean that the structural components of phenolic compounds are different in different cultivars of peach, because Yonemori et al. (1983) report that there are differences in the tannins of astringent and non-astringent fruits of Japanese persimmon (*Diospyros kaki*). As described later (see Tables 1 and 2), a large amount of total phenolic compounds in mature peach fruit resulted in a large amount of total soluble solids (Kubota et al. 1992). The cause is not known, but this relationship is of interest.

Table 2. Effects of girdling of lateral bearing branches at different stages of fruit growth on fruit weight, total soluble solids and phenolic contents at harvest, and PAL activity of Shimizu-hakuto peach fruit (Kubota et al. 1993)

Time of girdling[a]	Fruit weight (g)	Total soluble solids (°Brix)	Phenolic content (mg/100 g FW)	PAL activity[b] *t*-Cinnamic acid (nmol/g FW·h)	
				Girdled	Not girdled
Stage 1	242.2c[c]	8.6b	60.9b	218.0	65.0
Stage 2	362.8a	12.5a	77.4a	129.8	32.4
Stage 3	303.9b	11.4a	32.2c	0.6	0.3
Not girdled	274.8bc	10.0b	23.6c		

[a] Girdling was done on May 16 for stage 1, June 6 for stage 2, and July 4 for stage 3.
[b] Measured 7 days after girdling.
[c] Average values were analyzed with Duncan's multiple range test. Values within column followed by the same letter are not significantly different at $p = 0.05$.

3 PAL Activity in Peach Fruit

3.1 General Phenylpropanoid Metabolism

The biosynthetic pathways of phenolic compounds are understood in some detail (Swain 1965; Van Buren 1970; Rhodes et al. 1981; Haslam 1989; Macheix et al. 1990). The aromatic amino acids phenylalanine and tyrosine are produced in plants via the shikimate pathway. Phenylalanine is the most important precursor of phenolic compounds in higher plants except for some Gramineae such as *Oryza sativa*, and this amino acid leads to phenylpropanoid metabolism. The hydroxycinnamic acids are formed in this pathway, as are their coenzyme A (CoA) esters, which are common structural elements of various phenolic compounds, such as lignin, flavonoids, benzoic acids, cinnamate esters, and condensed tannins. In the transition from phenylalanine to *p*-coumaric acid, three enzymes, PAL, *trans*-cinnamate 4-monooxygenase (EC 1.14.13.11) and 4-coumarate CoA ligase (EC 6.2.1.12), are of particular importance in the biosynthesis of practically all phenolic compounds. The *trans*-cinnamate 4-monooxygenase and 4-coumarate CoA ligase make possible the conversion of cinnamic to 4- or *p*-hydroxycinnamic acid and the conversion of hydroxycinnamic acids to CoA esters, respectively.

PAL catalyzes the deamination of L-phenylalanine and produces *t*-cinnamic acid in the first step in phenolic metabolism, so PAL is referred to as the key enzyme or rate-limiting enzyme in phenolic biosynthesis. It is difficult to compare PAL activity in different circumstances, because of the large effects of the physiological stage on the enzyme and because of its sensitivity to external factors such as light, temperature and stress (Macheix et al. 1990).

3.2 PAL Extraction and Assay

3.2.1 Extraction and Purification

The method of Faragher and Chalmers (1977) can be used for extraction of PAL from plants. All chemicals are reagent-grade commercial products. Two g of fresh mesocarp tissue of peach fruit are ground in a mortar with 10 ml of 0.1 M borate buffer containing 20 mM 2-mercaptoethanol (pH 8.8), Polyclar AT (0.25 g/g of sample), and quartz sand (0.5 g/g of sample). A further 20 ml of buffer is added to the homogenate, and the mixture is centrifuged at 13 000 *g* for 30 min. To 20 ml of the supernatant, $(NH_4)_2SO_4$ is added to 70% saturation to precipitate the protein. The mixture is shaken vigorously and left for 30 min. After centrifugation of the mixture at 13 000 *g* for 15 min, the precipitate is suspended in 5 ml of the borate buffer and used as the crude enzyme solution. All steps are carried out below 4 °C.

Aoki et al. (1971) obtained the crude enzyme as an acetone powder. The procedure will be described here in brief. A fresh sample is blended with cold acetone (−20 °C). The homogenate is filtered and dried in a desiccator under reduced pressure at 0 °C overnight. The acetone powder can be stored in a freezer at −20 °C for at least 6 months with little loss of PAL activity. For preparation of the enzyme solution, 1 g of acetone powder is suspended in 40 ml of 0.05 M sodium borate buffer (pH 8.8) at approximately 3 °C for 1 h, and the suspension is cleared by centrifugation at 7000 *g* for 10 min at 0 °C. The supernatant obtained is used as the crude enzyme solution. The crude enzyme solution also is stable.

3.2.2 Spectrophotometric Assay

PAL activity is assayed spectrophotometrically based on the methods of Koukol and Conn (1961) and Kataoka et al. (1983) by measurement of the rate of formation of *t*-cinnamic acid as the increase in absorbance at 268 nm. The reaction mixture consists of 2 ml of the enzyme preparation and 1 ml of 25 μm L-phenylalanine. In the reference mixture, the L-phenylalanine solution is replaced by distilled water. The mixture is incubated at 37 °C for 3 h with shaking. The reaction is stopped by the addition of 0.1 ml of 6 N HCl. After 5 ml of peroxide-free ethyl ether is added, the mixture is shaken again vigorously for extraction of the acidified fraction and then centrifuged at 3000 *g* for 5 min. The ether phase is obtained with a pipette, and the ether is evaporated to dryness under reduced pressure. The residue is dissolved in 4 ml of 0.05 N NaOH and the absorbance of the solution at 268 nm in a 1-cm quartz cell is measured. The enzyme activity is expressed as micromoles of *t*-cinnamic acid formed per g FW or per mg protein in fresh tissue/h. The reaction product is scanned with a spectrophotometer for comparison of its optical characteristics with those of authentic *t*-cinnamic acid. The absorption spectrum of the reaction product was identical with that of authentic *t*-cinnamic acid (Fig. 3).

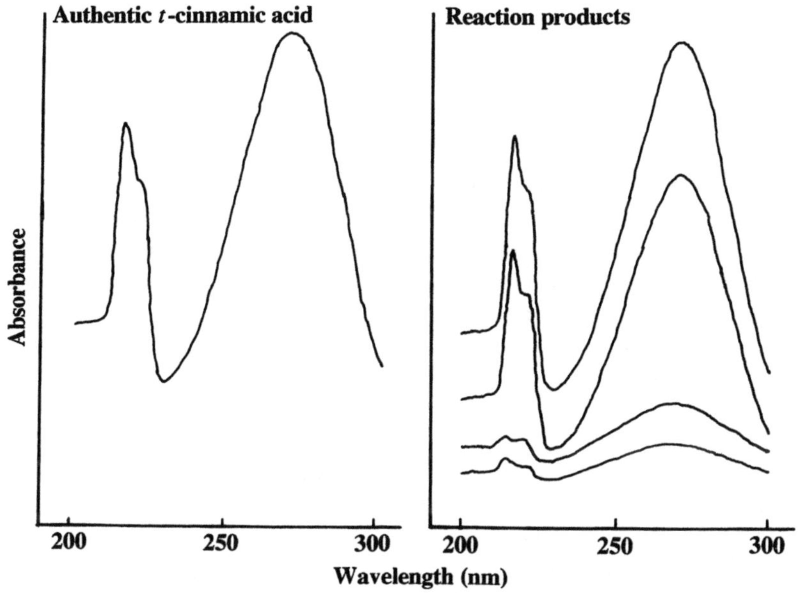

Fig. 3. Absorption spectra of authentic *t*-cinnamic acid (*left*) and the reaction products (*right*)

3.2.3 Radiochemical Assay with [^{14}C] Phenylalanine

PAL activity can be assayed by radiochemical analysis based on the methods of Dixon and Fuller (1976) and Hiramatsu et al. (1986). L-[U-^{14}C] phenylalanine (specific activity; 495 mCi/mM) can be purchased from Radiochemical Centre, Amersham, and other chemicals are reagent-grade commercial products. The reaction mixture, which consists of 100 µl of crude enzyme solution, 100 µl of 1 mM Tris-HCl buffer (pH 8.5), 1 µl of 1 mM dithiothreitol, and 25 µl of [^{14}C] phenylalanine (10 µCi/ml), is incubated at 30 °C for 1 h with shaking. The reaction is stopped by the addition of 500 µl of methanol. To the solution is added 2 ml of ethanol and the mixture is kept at 4 °C for more than 1 h for precipitation of proteinous substances. After centrifugation of the mixture at 3500 g for 10 min, the supernatant is concentrated to dryness under reduced pressure and then dissolved in a small amount of ethanol. The solution is spotted onto a TLC plate (Merck Art. 5735) and fractionated with a solvent system of toluene:ethyl formate:formic acid = 60:40:1 (v/v). After the plates are dried, the zone corresponding to acid (Rf of cinnamic acid, 0.51) is scanned under UV light, cut out, and put into vials with 500 µl of ethanol and 3 ml of scintillator (ACSII). The radioactivity incorporated into the cinnamic acid is counted with a liquid scintillation counter (Aloka LSC-653). The enzyme activity is expressed as disintegrations per minute of the ^{14}C incorporated into *t*-cinnamic acid on a FW basis or protein basis per hour as described above.

3.3 Relationships Between Phenolic Content, PAL Activity, and L-Phenylalanine Content

For evaluation of the relationship between the phenolic content and PAL activity in mesocarp tissue, changes during the growth of fruits of two peach cultivars, Mishima-hakuto and Shimizu-hakuto, were investigated by the methods summarized above. Changes in the concentration of L-phenylalanine, the substrate in the PAL reaction, were also measured, because enzyme activity is generally affected by the concentration of the substrate (Da Cunha 1987). Margna (1977) suggests that the biosynthetic rate of phenylpropanoid may be not controlled by PAL activity but by the L-phenylalanine concentration in plant cells. Amino acids including L-phenylalanine have been extracted from peach fruit with 70% (v/v) ethanol, fractionated on resin, and assayed with an automated amino acid analyzer. The results (unpubl.) are discussed below.

The growth curve of peach fruit as well as other stone fruits is double sigmoid, and it is generally classified into stages 1, 2 and 3 of fruit growth (Romani and Jennings 1971). Stage 1 of fruit growth results from the enlargement of all parts of the ovary other than the endocarp and embryo. During stage 2, lignification of the endocarp takes place and growth is confined principally to the endosperm and embryo. It is during stage 3 that expansion of the mesocarp tissue, leading to the mature fruit, resumes. The growth curve on a fruit weight basis was a double sigmoid with both cultivars tested (Fig. 4). Regardless of the cultivar, both the contents of total and HMWF (data not shown) phenols in the fruit decreased toward the end of stage 1, after which the phenolic contents increased rapidly and reached a maximum in the middle of stage 2. The content of phenolic compounds rapidly decreased to a lower lever during stage 3. These findings corresponded to those for other peach cultivars (Craft 1961; Goldstein and Swain 1963; Kubota et al. 1986). Throughout fruit growth, the contents of both total and HMWF phenols were higher in Mishima-hakuto fruit than in Shimizu-hakuto, especially during stage 2. The higher the total phenolic content in mature peach fruits, the higher the total phenolic content at the maximum, during stage 2 (Kubota et al. 1986).

Irrespective of the cultivar, PAL activity was high during stage 1 and reached a maximum in the middle of stage 2, after which no activity detected (Fig. 4). Throughout stage 1 and the first half of stage 2, PAL activity was higher in Mishima-hakuto fruit than in Shimizu-hakuto. This result may indicate that the increase in the total phenolic content during stage 2 is due to increased PAL activity during the early stages of fruit development. The L-phenylalanine contents, on the other hand, were high in the early stages of fruit growth irrespective of the cultivar, and then decreased and stayed low until maturation. There was no difference in the L-phenylalanine content between the two cultivars throughout fruit growth and maturation.

These findings suggested that the total phenolic content in peach fruit is closely related to the PAL activity. Further investigation is needed to identify the relationship between PAL activity and the L-phenylalanine concentration in peach fruit.

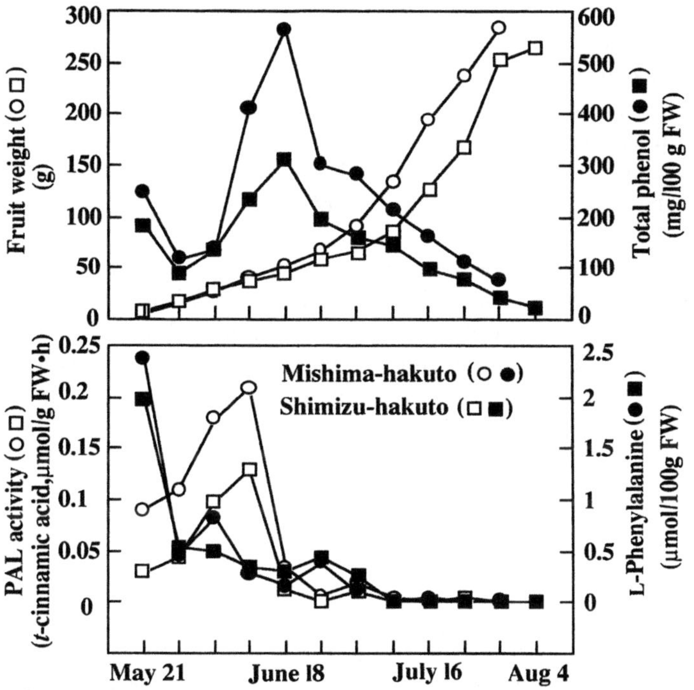

Fig. 4. Fruit growth curves of two peach cultivars, Mishima-hakuto and Shimizu-hakuto, and changes in the total phenolic content, PAL activity, and L-phenylalanine content of their mesocarp tissues

4 Factors Affecting Phenolic Content and PAL Activity in Peach Fruit

Phenolic content and PAL activity in plants are affected by various internal and external factors (Camm and Towers 1973; McClure 1979). For example, PAL activity is influenced by exogenous factors such light intensity, temperature, and the application of certain growth regulators, especially ethylene. Unlike grape and apple fruits (Faragher 1983; Roubelakis-Angelakis and Kliewer 1986), there is little information about the effects of these factors on phenolic content and PAL activity in peach fruit. Based on our experimental results, the relationship between these parameters in peach fruit is discussed below.

4.1 Inherent Factors

The contents of phenolic compounds in mature peach fruit varied widely among the cultivars, sometimes with a difference of eightfold (unpubl.). For instance, the levels of phenolic compounds in mature fruit were compared among 20 peach (var.

vulgaris Maxim.) and three nectarine (var. *nucipersica* Schneid.) cultivars, and the total phenolic content of the fruit of Wase-momoyama, Yahata-hakuho, Shimizu-hakuto and Setouchi-hakuto was from 20.5 to 35.9 mg/100 g FW, but the fruit of Mishima-hakuto, Hakuto, and Golden Peach contained from 80.0 to 161.4 mg phenolics. The content of HMWF was higher in the latter cultivars than in the former ones. In the nectarine fruit, the total phenolic content was high, from 67.0 to 150.4 mg, in all cultivars tested, and the amount of HMWF phenolics was also high.

Moreover, the phenolic content varied among the different parts of the fruit, with more in the skin (peel) and the seed than in mesocarp tissue (Kubota et al. 1992). Kataoka et al. (1983) found that PAL activity in the skin and seeds of grape berries was higher than that in the pulp during the early stages of berry growth. It seems that the high content of phenolic compounds in the skin and the seed of fruits are related to coloration and lignification, respectively.

4.2 Environmental Factors

Of all environmental factors, light intensity is the most important one influencing PAL activity. When exposure of peach fruit to light is limited by bagging of the fruit, the total phenolic content in the mesocarp tissue of mature fruit is lower than without bagging (Kubota et al. 1992). Temperatures below 20 °C during fruit growth, especially in stage 3, increase the total phenolic content (Kubota et al. 1990). Hesse (1975) suggests that astringency of peach fruit is caused by low temperatures during fruit growth.

The soil moisture content affects the phenolic content of peach fruit (Kubota and Kudo 1992). In our experiment, pot-grown peach trees were subjected to a soil moisture tension, pF, of 2.3 to 2.7 by withholding of water during one of the three stages of fruit growth, and the pF of control trees was kept at 1.5 or below throughout the experiment. Trees from which water was withheld during stage 3 of fruit growth had not only the smallest fruits with the largest amount of soluble solids, but also a higher phenolic content in the fruit than the control trees. However, no difference was observed in PAL activity in fruit from the trees stressed during stage 3 and the control trees. In addition, when peach trees were kept at one of four ranges of soil moisture tension (below 1.5, for the control, 1.5 to 2.0, 2.0 to 2.5, and above 2.5) while the fruit was in stage 3, the total phenolic content of fruit harvested from trees held at a pF of 2.0 or higher was twice that of the trees kept at a pF of 2 or less, which included the control trees (Table 1). The higher pFs resulted in higher PAL activity by the radiochemical assay, but by the spectrophotometric assay, PAL activity was not detectable in any fruits. These observations may indicate that soil moisture tension above a pF 2 during stage 3 of fruit growth is closely related to the occurrence of astringency in peach fruit.

4.3 Other Factors

The contents of phenolic compounds and PAL activity in peach fruit are affected by the rootstock used. The content of phenolic compounds in fruit from trees grafted on dwarfing stocks such as Nanking cherry (*P. tomentosa*) or Chinese bush berry (*P. japonica*) and in fruit from trees grafted on the usual type, Juseito (*P. persica*), during fruit maturation, has been compared (Kubota et al. 1986). Throughout the growth of fruit, the total phenolic content was higher in fruit from trees grafted on Nanking cherry or Chinese bush berry rootstock than in those from trees grafted on Juseito stock, the content was especially great with Chinese bush berry. PAL activity was higher in fruit from trees with Nanking cherry roostock throughout stage 1 and the first half to stage 2.

A large amount of phenolic compounds in peach fruit has been observed in trees subjected to stress such as girdling. We (Kubota et al. 1993) observed that when the lateral bearing branches of a peach tree are artificially girdled 1 cm wide at stages 1, 2 or 3 of fruit growth, the phenolic content of mature fruit from branches treated at stage 1 or 2 is higher than that in fruit from branches girdled at stage 3 or notgirdled (controls; Table 2). The PAL activity assayed 7 days after girdling by spectrophotometry was higher in fruit from branches girdled at stages 1 or 2 than in fruit from control branches or those girdled at stage 3.

5 Concluding Remarks

In summary, results obtained from various studies indicate that: (1) the total phenolic contents of mature peach fruit vary widely among cultivars; (2) regardless of the cultivar, the phenolic content reaches a maximum in stage 2 of fruit development, and then decrease rapidly until harvest; (3) PAL activity is higher during stage 1 and the first half of stage 2 than later, and the higher the content of phenolic compounds in peach fruit, the higher the PAL activity; (4) some exogenous factors such as soil moisture stress, low temperature, strong light intensity, and girdling of branches result in a high content of phenolic compounds and presumably high PAL activity in peach fruit, although the degree of the effects vary in the various stages of fruit development.

The phenolic content is related to the PAL activity in the mesocarp tissue of peach fruit. However, soil drought during stage 3 of fruit growth, in which generally no PAL activity is detectable, increases the content of phenolic compounds, although PAL activity in fruit from water-stressed trees is slightly higher than those from the control trees. This observation suggests that other mechanisms may act to increase the phenolic content, presumably through inhibition of degradation of phenolic compounds. In addition, nectarine fruits contain a large amount of phenolic compounds, although the fruits have almost no astringency.

This finding may mean that the structural components of phenolic compounds in nectarine fruit are different from those in peach fruit.

References

Aoki S, Araki C, Kaneko K, Katayama O (1971) Occurrence of L-phenylalanine ammonia-lyase activity in peach fruit during growth. Agric Biol Chem 35:784–787

Bate-Smith EC (1973) Haemanalysis of tannins: the concept of relative astringency. Phytochemistry 12:907–912

Camm EL. Towers GHN (1973) Phenylalanine ammonia lyase. Phytochemistry 12:961–973

Craft CC (1961) Polyphenolic compounds in Elberta peaches during storage and ripening. Proc Am Soc Hortic Sci 78:119–131

Da Cunha A (1987) The estimation of L-phenylalanine ammonia-lyase shows phenylpropanoid biosynthesis to be regulated by L-phenylalanine supply and availability. Phytochemistry 26:2723–2727

Dixon RA, Fuller KW (1976) Effects of synthetic auxin levels on phaseollin production and phenylalanine ammonia-lyase (PAL) activity in tissue cultures of *Phaseolus vulgaris* L. Physiol Plant Pathol 9:299–312

Faragher JD (1983) Temperature regulation of anthocyanin accumulation in apple skin. J Exp Bot 34:1291–1298

Faragher JD, Chalmers DJ (1977) Regulation of anthocyanin synthesis in apple skin. III. Involvement of phenylalanine ammonia-lyase. Aust J Plant Physiol 4:133–141

Goldstein JL, Swain T (1963) Changes in tannins in ripening fruits. Phytochemistry 2:371–383

Guadagni DG, Nimmo CC (1953) Effect of growing area on tannin and its relation to astringency in frozen Elberta peaches. Food Technol 7:59–61

Hagerman AE (1989) Chemistry of tannin-protein complexation. In: Hemingway RW, Karchesy JJ (eds) Chemistry and significance of condensed tannins. Plenum Press, New York, pp 323–333

Hartley RD (1987) HPLC for the separation and determination of phenolic compounds in plant cell walls. In: Linskens HF, Jackson JF (eds) Modern methods of plant analysis, vol 5. High performance liquid chromatography in plant sciences. Springer, Berlin Heidelberg New York, pp 92–103

Haslam E (189) Plant polyphenols:vegetable tannins revisited. Cambridge University Press, Cambridge, pp 195–219

Hesse CO (1975) Peaches. In: Janick J, Moore JN (eds) Advances in fruit breeding. Purdue University Press, West Lafayette, IN, pp 285–335

Hiramatsu M, Ichinose Y, Shiraishi T, Oku H, Ouchi S (1986) Regulation of pisatin biosynthesis in pea leaves by elicitor and suppressor produced by *Mycosphaerella pinodes*. Ann Phytopathol Soc Jpn 52:53–58

Joslyn MA, Goldstein JL (1964) Astringency of fruits and fruit products in relation to phenolic content. Adv Food Res 13:179–217

Karchesy JJ (1989) Analytical methods: an overview. In: Hemingway RW, Karchesy JJ (eds) Chemistry and significance of condensed tannins. Plenum Press, New York, pp 197–204

Kataoka, I, Kubo Y, Sugiura A, Tomana T (1983) Changes in L-phenylalanine ammonia-lyase activity and anthocyanin synthesis during berry ripening of three grape cultivars. J Jpn Soc Hortic Sci 52:273–279

Koukol J, Conn EE (1961) The metabolism of aromatic compounds in higher plants. IV. Purification and properties of the phenylalanine deaminase of *Hordeum vulgare*. J Biol Chem 236:2692–2698

Kubota N, Kudo S (1992) Effects of soil moisture tension on phenolic contents and astringency in peach fruits. J Jpn Soc Hortic Sci 61:31–37 (in Japanese with English Summary)

Kubota N, Shimamura K, Mimura H, Nakano M (1986) Changes in the content of total phenolic compound during fruit ripening of peaches as affected by different rootstocks. HortScience 21:765 (Abstr 774)

Kubota N, Kawao T, Shimamura K (1990) Effects of soil drought and air temperature on the contents of phenolic compound in relation to astringency of peach fruits. Environ Control Biol 28:141–146 (in Japanese with English Summary)

Kubota N, Mimura H, Yakushiji H, Shimamura K (1992) Astringency of peach fruit in different fruit parts, trees and orchards (in Japanese with English summary). Sci Rep Fac Agric Okayama Univ 79:45–51

Kubota N, Takagi S, Kudo S (1993) Phenolic contents in peach fruits as influenced by tree vigor and girdling of scaffold limbs. J Jpn Soc Hortic Sci 62:83–88 (in Japanese with English Summary)

Macheix J-J, Fleuriet A, Billot J (1990) Fruit phenolics. CRC Press, Boca Raton

Margna U (1977) Control at the level of substrate supply: an alternative in the regulation of phenylpropanoid accumulation in plant cells. Phytochemistry 16:419–426

McClure JW (1979) The physiology of phenolic cmpounds in plants. In: Swain T, Harbone JB, Van Sumere CF (eds) Biochemistry of plant phenolics. Plenum Press, New York, pp 525–556

Nakabayashi T (1968) Studies on tannin of fruits and vegetables. Part 1. Tannin of fruits of rosaceous fruit-trees. J Jpn Soc Food Sci Technol 15:73–78 (in Japanese with English Summary)

Okuda T, Mori K, Hatano T (1985) Relationship of the structures of tannins to the binding activities with hemoglobin and methylene blue. Chem Pharm Bull 33:1424–1433

Ozawa T, Lilley TH, Haslam E (1987) Polyphenol interactions: astringency and the loss of astringency in ripening fruit. Phytochemistry 26:2937–2942

Porter LJ, Woodruffe J (1984) Haemanalysis: the relative astringency of proanthocyanidin polymers. Phytochemistry 23:1255–1256

Reeve RM (1959) Histological and histochemical changes in developing and ripening peaches. I. The catechol tannins. Am J Bot 46:210–217

Rhodes MJC, Wooltorton LSC, Hill AC (1981) Changes in phenolic metabolism in fruit and vegetable tissues under stress. In: Friend J, Rhodes MJC (eds) Recent advances in the biochemistry of fruits and vegetables, Academic Press, London, pp 193–220

Romani RJ, Jennings WG (1971) Stone fruits. In: Hulme AC (ed) The biochemistry of fruits and their products, vol 2. Academic Press, London, pp 411–436

Roubelakis-Angelakis KA, Kliewer WM (1986) Effects of exogenous factors on phenylalanine ammonia-lyase activity and accumulation of anthocyanins and total phenolics in grape berries. Am J Enol Vitic 37:275–280

Schwarzenbach R (1982) High-performance liquid chromatography of carboxylic acids. J Chromatogr 251:339–358

Senter SD, Callahan A (1990) Variability in the quantities of condensed tannins and other major phenols in peach fruit during maturation. J Food Sci 55:1585–1587, 1602

Sistrunk Wa (1985) Peach quality assessment: fresh and processed. In: Pattee HE (ed) Evaluation of quality of fruits and vegetables. AVI Publishing, Westport, CT, pp 1–46

Swain T (1965) The tannins. In: Bonner J, Varner J (eds) Plant biochemistry. Academic Press, New York, pp 552–580

Swain T, Hillis WE (1959) The phenolic constituents of *Prunus domestica*. 1. The quantitative analysis of phenolic constituents. J Sci Food Agric 10:63–68

Van Buren J (1970) Fruit phenolics. In: Hulme AC (ed) The biochemistry of fruits and their products. vol 2. Academic Press, London, pp 269–304

Yonemori K, Matsushima S, Sugiura A (1983) Differences in tannins of non-astringent and astringent type fruits of Japanese persimmon (*Diospyros kaki* Thunb.). J Jpn Soc Hortic Sci 52:135–144 (in Japanese with English Summary)

Astringency in Persimmon

S. Taira

1 Introduction

The genus *Diospyros* (family, Ebanaceae), to which persimmons belong, contains about 400 species, most of which are found in subtropical to tropical regions. The wood from certain species of the genus is used for furniture and the heads of golf clubs. For fruit production, only four species, *D. kaki* L., *D. lotus* L., *D. virginiana* L. and *D. oleifera* Cheng, are important (Kitagawa and Glucina 1984). *D. kaki* is also referred to as kaki (a word of Japanese origin meaning persimmon). It is the most important species, and the fruits are consumed both fresh and dried. The other three species are used mainly as a rootstock for persimmon or a source of tannins.

Persimmon seems to have originated in China, and has been cultivated in China, Korea and Japan for hundreds of years. Numerous cultivars, probably 1000 or more, with a wide variety of fruit sizes, shapes and colours, have been produced in Japan. Some have already disappeared. Nowadays in Japan, persimmon is the fifth most widely consumed fruit, but there are only a few cultivars that are commercially important. In recent years, interest in persimmon has been increasing worldwide, and a few cultivars have been introduced and cultivated elsewhere, including Brazil, Italy, the USA, Israel, New Zealand and Australia.

In pomology, persimmon fruits are often classified into four types, depending on their astringency at harvest maturity and on the degree of flesh browning caused by pollination, or more precisely by seed formation (Ito 1971; Itoo 1986): (1) pollination constant and non-astringent (PCNA), which is non-astringent whether or not there are seeds; (2) pollination constant and astringent (PCA), which is astringent whether or not there are seeds; (3) pollination variant and non-astringent (PVNA), which is astringent if there are few or no seeds; and (4) pollination variant and astringent (PVA), which has non-astringent flesh around the seeds only. Generally, we call PCNA and PVNA fruits sweet persimmons because they are non-astringent and sweet at harvest maturity, and call PCA and PVA fruits astringent persimmons.

Fruits of all types are strongly astringent when small and immature. PCNA and PVNA fruits lose astringency during development on the tree and become non-astringent by harvest with firm flesh, but the PCA and PVA fruits, being astringent even then, are inedible even when fully coloured. They lose astringency

Modern Methods of Plant Analysis, Vol. 18
Fruit Analysis
Edited by H.-F. Linskens and J.F. Jackson

when they become over-ripe with extremely soft flesh. Flesh browning is observed in PVNA and PVA persimmon fruit when pollination has been successful and seeds have formed. In PVNA fruit, those with few seeds fail to lose astringency completely, parts of the flesh remaining astringent. Such fruit cannot be eaten when firm unless the astringency has been removed artificially. Post-harvest carbon dioxide gas and alcohol (ethanol) vapour treatments are well-known methods for removal of astringency.

The strongly astringent taste of persimmon fruits arises from soluble tannins that accumulate in large specialized cells called tannin cells. In this chapter, some chemical properties of persimmon tannins, an outline of simple and reliable methods for estimation of the degree of astringency and the amount of tannins, and some aspects of the physiological changes in the tannin contents of persimmon fruits will be described.

2 Persimmon Tannin

Tannins are plant polyphenols of high molecular weight that can be used to tan animal skins in making leather. Persimmon tannins are of particularly high molecular weight, and the mixture contains enough phenolic hydroxyl groups to cause an astringent (drying and puckering) sensation in the mouth. This sensation is thought to be due to dehydration following the interaction of tannins with proteins on the surface of the tongue (Van Buren 1970; Kitagawa and Glucina 1984; Deshpande et al. 1986; Macheix et al. 1990). When an astringent persimmon fruit is eaten, the tannin cells in the flesh are crushed and soluble tannins are released, giving a strongly astringent sensation.

In Japan, crude persimmon tannin extracts called kaki-shibu (shibu means astringency) were used in the past to waterproof paper used in traditional umbrellas and in one kind of fan; the extract was used also to treat fibrous nets for fishing to make them stronger. Now, kaki-shibu is widely used as a deproteinizing agent in the brewing of sake (Japanese rice wine). Persimmon tannins have been suggested to have physiological effects such as the reduction of high blood pressure and also antibacterial effects.

Thorough reviews of fruit phenolics have appeared recently (Deshpande et al. 1986; Macheix et al. 1990), so only a brief history of the identification of persimmon tannins will be given here, together with some qualitative differences among the tannins in different fruit types.

2.1 Chemical Structure

Early chemical studies of persimmon tannin were often done with insufficiently purified tannin fractions. It was reported in 1923 that the elemental formula of persimmon tannin might be $C_{14}H_{20}O_9$, and in 1962, it was proposed that the major component of persimmon tannin was leucodelphinidin-3-glucoside. Further studies done by Itoo et al. showed that persimmon tannin might have a more complex

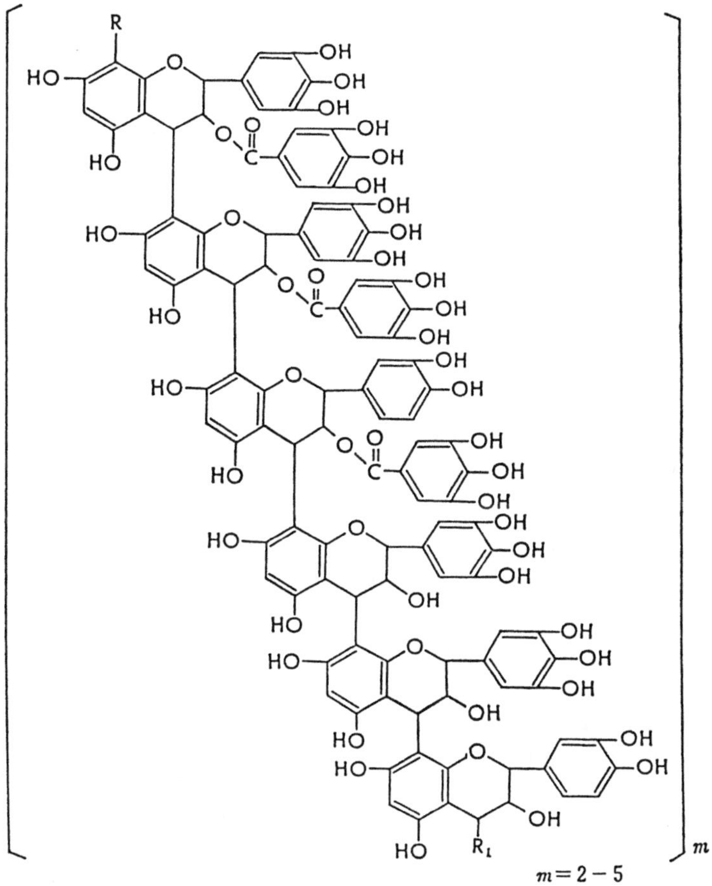

Fig. 1. Presumed repeat in the structure of persimmon tannin. (Matsuo and Ito 1977b)

structure. These workers suggested that persimmon tannin is a kind of conjugated tannin, whose major component is leucodelphinidin, to which gallic acid, gallocatechin and gallocatechin gallate are conjugated (Matsuo and Ito 1978; Itoo 1986). However, the chemical structure remained unclear, and the molecular weight could not be estimated closely. Despite much study, the precise chemical structure of persimmon tannin remained unidentified.

Matsuo and Ito (1977b, 1978) succeeded in purifying persimmon tannin in more detail than before and suggested that it consisted of catechin, catechin-3-gallate, gallocatechin and gallocatechin-3-gallate in the ratio of 1:1:2:2 plus an unknown terminal residue. Estimations of the molecular weight of intact persimmon tannin are 1.12×10^4 (M_w) and 0.52×10^4 (M_n). The presumed repeating structure of persimmon tannin is shown in Fig. 1. Piretti et al. (1985) pursued supplementary research to obtain similar results, with a slight difference in the

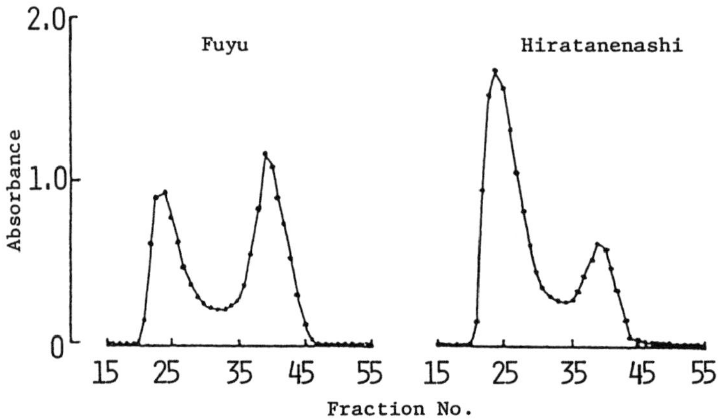

Fig. 2. Differences in separation pattern of soluble tannins from Fuyu (non-astringent) and Hiratanenashi (astringent) persimmon. (Yonemori and Matsushima 1984)

ratios of the four constituents named above. The researchers suggested that the difference probably arose because fruits of a different cultivar were used.

2.2 Differences Among Fruit Types

The basic structure of persimmon tannin seems to be, as proposed by Matsuo and Ito (1977b, 1978), a proanthocyanidin polymer. Some differences in the constituents and chemical properties of persimmon tannin among the four fruit types as classified in terms of astringency have been reported. Nakabayashi (1971) found three patterns of low-molecular-weight polyphenols (one in PCNA fruit, one in PVNA fruit and one in PVA and PCA fruit); the components were different. Yonemori et al. (1983) reported differences observed in the amounts of catechin and gallic acid in tannins extracted from immature PCNA fruit and the immature fruit of the three other types.

The differences in the chemical properties of persimmon tannins between immature PCNA fruit of the cultivar Fuyu and immature PVA fruit of the cultivar Hiratanenashi were further investigated by exclusion chromatography (Yonemori and Matsushima 1984). Tannins from the former cultivar were generally of lower molecular weight than those from the latter (Fig. 2). Fuyu tannins tended to coagulate more slowly in the presence of acetaldehyde vapour than those from Hiratanenashi, indicating that polymerization occurred less readily.

3 Analytical Methods

Often, rather than direct measurement of the level of tannins in samples of plant tissues, the amount of phenolic compounds is measured. Methods to measure phenolic substances are based mostly on the polarity of the phenolic hydroxyl groups, which have strong reducing power. The most common methods for tannin measurement include the vanillin assay, Lowenthal method (potassium permanganate titration), Folin-Denis method, ferric chloride reaction, Prussian blue assay and protein precipitation (Deshpande et al. 1986). Here, a few methods commonly used to estimate the levels of persimmon tannin are outlined.

3.1 Tannin Print Method

Persimmon tannin in soluble form reacts with ferric chloride, forming tannin-Fe ion complexes, which are blue-black. We can estimate the degree of astringency in persimmon fruits by colour development during the reaction.

Persimmon fruits are cut longitudinally and the freshly cut surface is pressed against filter paper that had been previously soaked in 5% ferric chloride solution and then dried. This method is called the tannin print method (Eaks 1967; Matsuo and Ito 1977a). There is a close relationship between the colour development of such prints and the amount of soluble tannin or the degree of astringency when the fruit is tasted (Sugiura et al. 1975, 1977). Figure 3 shows typical tannin prints of fruit at five stages of post-harvest treatment with ethanol vapour for removal of astringency in Hiratanenashi persimmon.

3.2 Folin-Denis Method

Among the methods for the measurement of soluble persimmon tannins, the Folin-Denis method is one of the most widely used. This colorimetric method is based on the reduction of Folin-Denis reagent (phosphomolybdic-phosphotungstic reagent) by soluble tannins in alkaline solution. The phenol (Folin-Ciocalteu) reagent, which has stronger sensitivity to reduction than the Folin-Denis reagent, is often used instead (Deshpande et al. 1986). Swain and Hillis (1959) proposed a modified method, which came to be much used in the routine analysis of large numbers of specimens. Figure 4 shows an outline of the procedure that the present author usually uses for the assay of soluble tannin in persimmon fruits.

Phenolic compounds such as persimmon tannin are generally soluble in polar solvents, so methanol or ethanol is often used as the extraction solvent. For the extraction of persimmon tannin, the most effective concentration of these solvents in aqueous solution is about 70% (Kato 1984). The sample should not contain more than 0.5 ml of methanol or ethanol because otherwise a cloudy white precipitate will form during the reaction with phenol reagent. The blank contains only water and the reagents. Calibration curves made with gallic acid and (+)-catechin are shown in Fig. 5. Gallic acid might be the better standard for calibration because the number of crystal water molecules in (+)-catechin is variable. The threshold for the sensation of astringency with the Folin-Denis method seems to be about

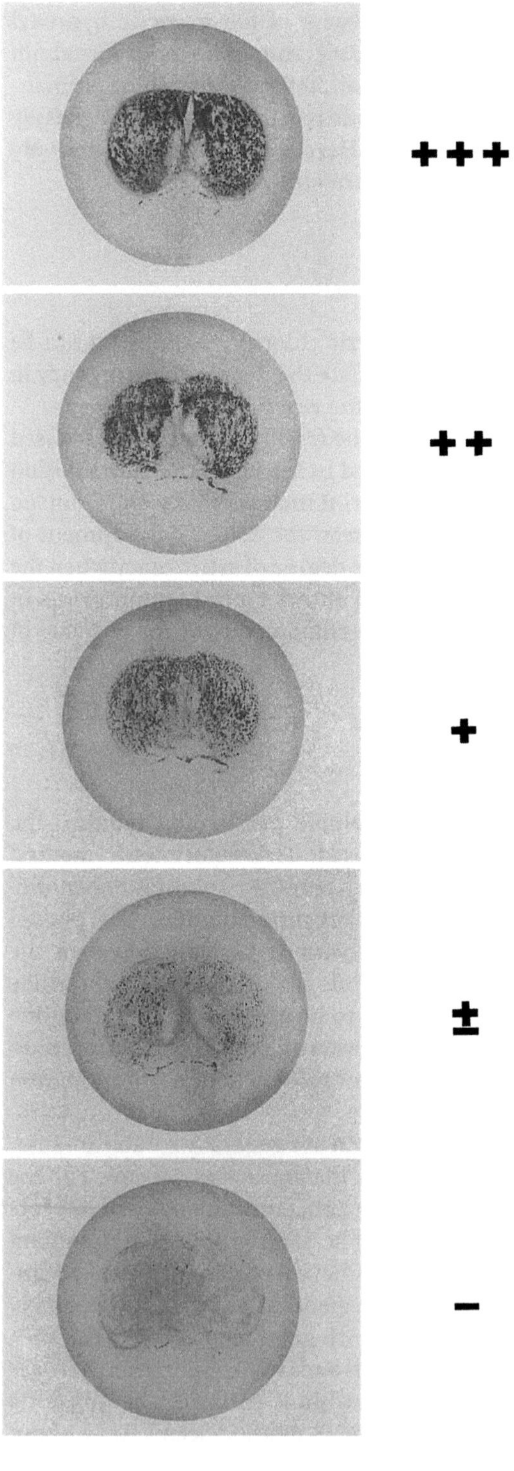

Fig. 3. Changes in results seen in tannin prints of freshly cut surfaces of Hiratanenashi persimmon at five stages of post-harvest treatment with ethanol vapour for removal of astringency. *Top* Before treatment; *bottom* at the end of treatment; +++ strongly astringent; − not astringent (by sensory evaluation)

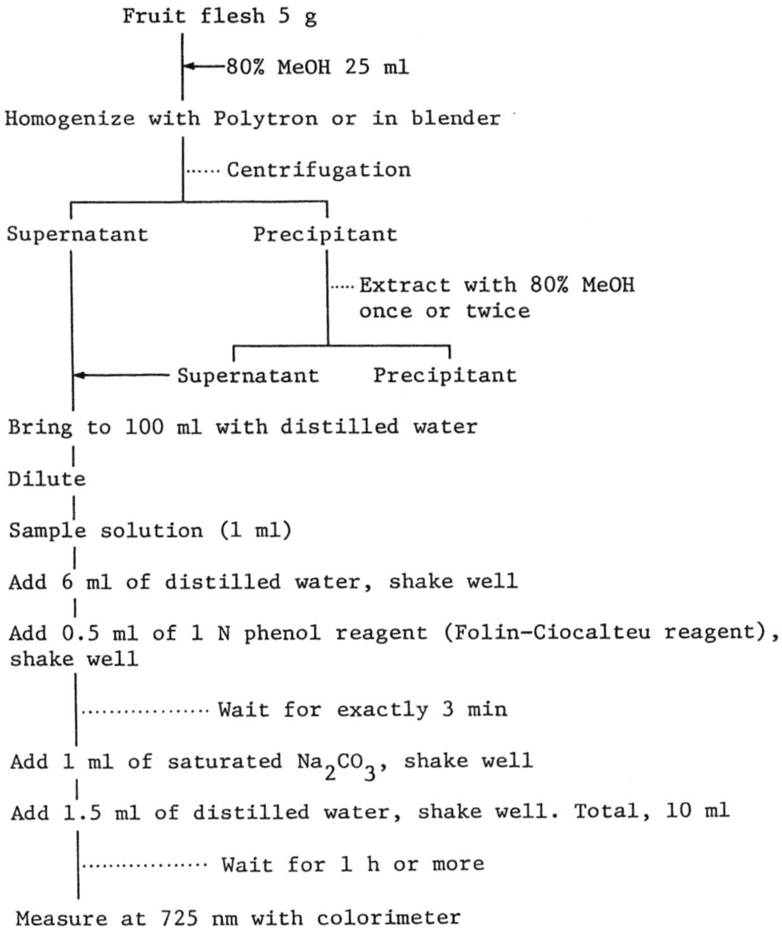

Fig. 4. Procedure for the measurement of soluble tannins (Folin-Denis method)

0.1% of soluble tannin on a fresh weight basis. Fruits containing less than this amount are almost non-astringent when eaten (Kato 1984; Ben-Arie et al. 1991).

The major drawback of the Folin-Denis method is that the reagent reacts with other phenolic constituents such as xanthine, proteins and amino acids. However, persimmon flesh usually contains small amounts of such substances, so the values obtained by the method probably reflect the amount of soluble tannins present.

3.3 Analysis of Total Tannins

During natural or artificial removal of astringency, soluble tannin can change, probably by polymerization or conjugation, into an insoluble form, which is not

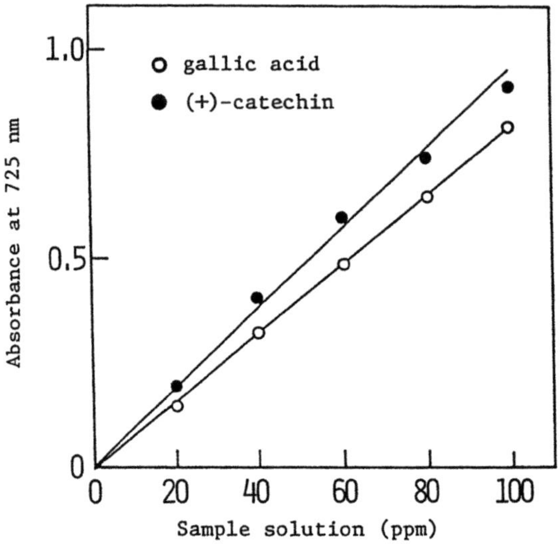

Fig. 5. Calibration curves of gallic acid and (+)-catechin measured by the Folin-Denis method

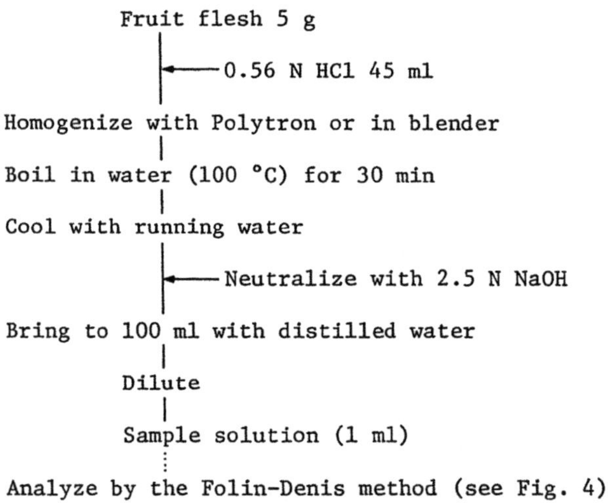

Fig. 6. Procedure for the measurement of total tannins

astringent (Ito 1971; Kitagawa and Glucina 1984). Thus, the changes in the amounts of total (soluble + insoluble) or insoluble tannin during fruit development of non-astringent and astringent persimmons are of physiological interest in terms of the ability of the persimmons to synthesize tannin. Knowledge about changes in the amount of total tannin would make the mechanisms of astringency removal more clear.

The hydrolysis of insoluble tannin with HCl solution can be used to measure the amount of total tannin (Manabe 1982). After insoluble tannins are hydrolyzed as shown in Fig. 6, total tannins can be measured by the Folin-Denis method. My group used this method in studying both PCNA and PVA fruit (Taira et al. 1990).

Insoluble tannins that formed in the PVA fruit during post-harvest treatment for removal of astringency could be extracted, but those that formed in the PCNA fruit during fruit development could not be readily extracted, though the fruits were astringent when immature. This finding suggested that the insoluble tannins that formed in astringent and non-astringent fruits were different.

4 Quantitative Changes in Tannins

Persimmon fruits characteristically accumulate high levels of soluble tannins during development. The accumulation starts in the ovary in anthesis, that is, in an extremely early stage of fruit development. Soluble tannins in non-astringent fruit decrease spontaneously to only trace amounts or can change into insoluble forms during fruit development on the tree, and such tannins in astringent fruits can be changed during treatment for removal of astringency. Here, changes in the amount of soluble or total tannin during fruit development and during certain treatments for removal of astringency studied recently by my group (Taira et al. 1990) will be described.

4.1 Fruit Development

Seasonal changes in fruit diameter, colour development of the peel (estimated from colour charts) and soluble tannins of persimmon fruits of the cultivars Hiratanenashi (PVA, a parthenocarpic type; Fig. 7A) and Jiro (PCNA; Fig. 7B) were studied ca.every 10 days at Tsuruoka City in the north of Japan. Soluble tannins in the fruit flesh were measured by the Folin-Denis method. In Tsuruoka, both cultivars usually bloom, starting early in June.

Figures 8 and 9 show the results for Hiratanenashi and Jiro fruits, respectively. The growth curves of the fruits were double-sigmoidal in shape. The fruits of Hiratanenashi and Jiro reached harvest maturity in late October or the middle of November, respectively. In Hiratanenashi, the soluble tannin concentration in the flesh gradually declined during fruit development, and was about 1.0% when the fruit reached full maturity. In Jiro, the soluble tannin concentration decreased rapidly, starting in the middle of June, and the fruit became non-astringent in September. The amount of soluble tannins per fruit increased as the fruit developed and reached a maximum at the end of the second growth stage (slow enlargement phase) in both cultivars. The amount of soluble tannins that accumulated in Jiro fruit was much less than that in Hiratanenashi fruit.

Almost all of the tannin in the fruit of astringent Hiratanenashi seems to be in soluble form (Taira et al. 1990). The amount of insoluble tannins in non-astringent Jiro fruits could not be measured, due to difficulty in the extraction of the insoluble tannin (Taira et al., unpubl.).

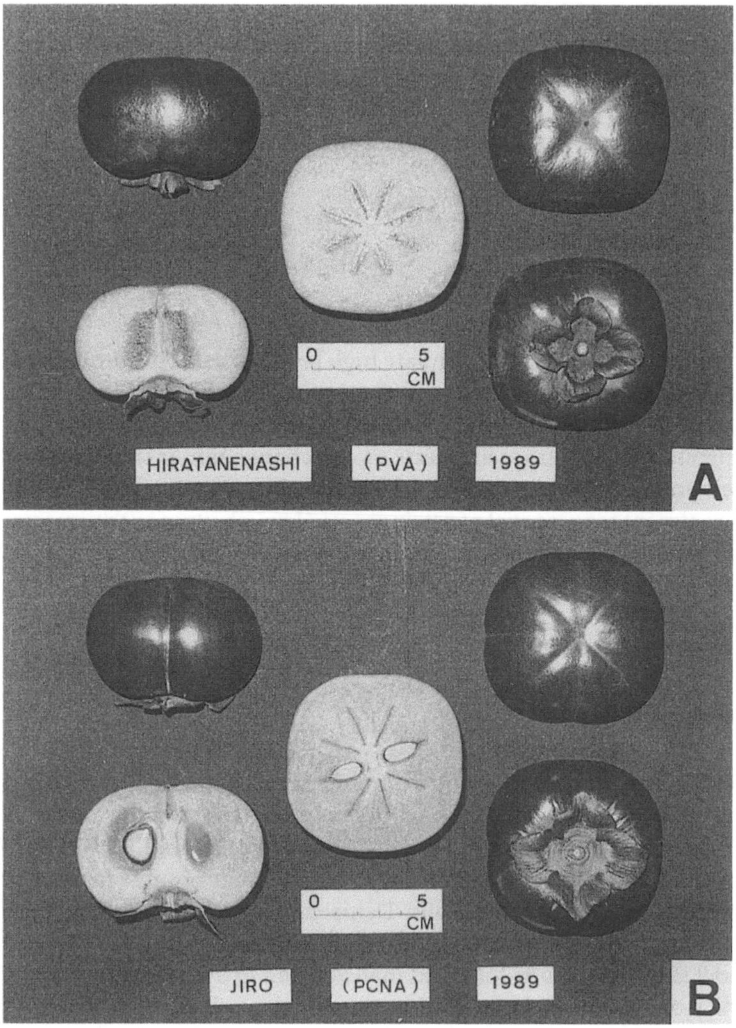

Fig. 7A,B. Fruits of persimmon. ***A*** Hiratanenashi; astringent and parthenocarpic. ***B*** Jiro; non-astringent

4.2 Removal of Astringency

Astringency can be removed from astringent persimmon fruits with various treatments, including ethanol vapour treatment of fruit on the tree or after harvest, carbon dioxide gas treatment, immersion into warm water, drying of peeled fruit, freezing of fruit and fruit softening or over-ripening on the tree or after harvest (Kitagawa and Glucina 1984). Astringency is removed by soluble tannins becoming insoluble, mainly because of the polymerization or condensation with acetaldehyde produced in the fruit flesh during the treatments listed except for

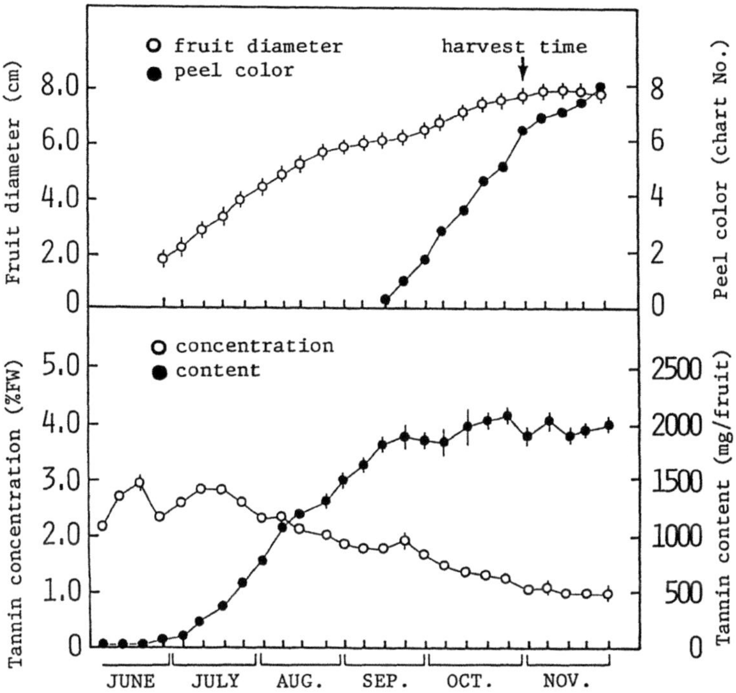

Fig. 8. Changes in fruit diameter, peel colour, and soluble tannins of the flesh during fruit development in Hiratanenashi

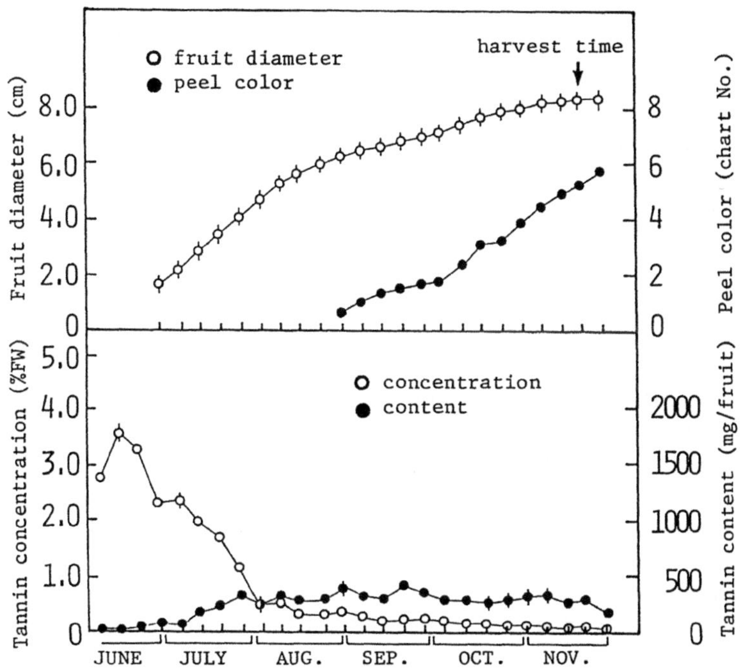

Fig. 9. Changes in fruit diameter, peel colour, and soluble tannins of the flesh during fruit development in Jiro

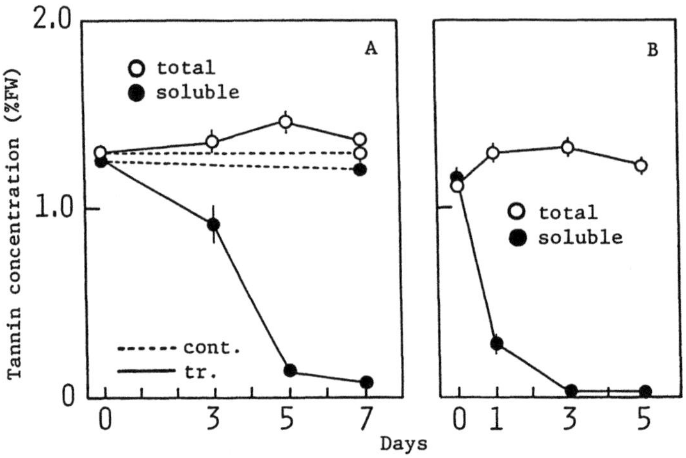

Fig. 10A,B. Changes in soluble and total tannins during post-harvest treatments for removal of astringency in Hiratanenashi persimmon. ***A*** Ethanol vapour treatment. ***B*** Carbon dioxide gas treatment

fruit freezing (Ito 1971; Itoo 1986). The mechanism of the decrease in astringency during freezing is unknown.

Astringent Hiratanenashi fruits were treated for removal of astringency with ethanol vapour or carbon dioxide gas after harvest, and the changes in soluble and total tannins in the fruit flesh were assayed by the methods of Folin-Denis and Manabe, respectively. The soluble tannin concentrations decreased during the treatments, and the fruits became non-astringent in taste in 7 and 3 days for ethanol vapour and carbon dioxide gas treatment, respectively (Fig. 10). Total tannin concentrations were almost constant and fruit fresh weight did not change during either treatment, which findings showed that the soluble tannins became insoluble during treatment.

5 Notes on Analytical Methods

As mentioned before, the amount of soluble tannin that causes astringency in persimmon fruits is usually estimated visually by the tannin print method and can be measured quantitatively by the Folin-Denis method. There is also a protein precipitation method for the measurement of soluble tannins (Hagerman and Butler 1978). In that method, the soluble tannin content is assayed by the addition of the sample to a standard solution of protein and the isolation of insoluble tannin-protein complexes. The complexes are dissolved in alkaline solution, to which ferric chloride is added. The absorbance of the solution at 510 nm is measured.

The method of Manabe (1982) for total tannin determination works well with astringent fruits such as Hiratanenashi, but not with non-astringent fruits such as

Jiro. The method for the extraction of insoluble tannins from the fruits of non-astringent (PCNA) persimmon needs improvement, if tannin biosynthesis during fruit development is to be studied further.

References

Ben-Arie R, Zutkhi Y, Sonego L, Klein J (1991) Modified atmosphere packaging for long-term storage of astringent persimmons. Postharvest Biol Technol 1:169–179

Deshpande SS, Cheryan M, Salunkhe DK (1986) Tannin analysis of food products. CRC Crit Rev Food Sci Nutr 24:401–449

Eaks IL (1967) Ripening and astringency removal in persimmon fruits. Proc Am Soc Hortic Sci 91:868–875

Hagerman AE, Butler LG (1978) Protein precipitation method for the quantitative determination of tannins. J Agric Food Chem 26:809–812

Ito S (1971) The persimmon. In: Hulme Ac (ed) The biochemistry of fruits and their products, vol 2. Academic Press, London, pp 281–301

Itoo S (1986) Persimmon. In: Monselise SP (ed) CRC handbook of fruit set and development. CRC Press, Boca Raton, pp 355–370

Kato K (1984) The condition of tannin and sugar extraction, the relation of tannin concentration to astringency and the behaviour of ethanol during the de-astringency by ethanol in persimmon fruits. J Jpn Soc Hortic Sci 53:127–134 (in Japanese with English Summary)

Kitagawa H, Glucina PG (1984) Persimmon culture in New Zealand. Science Information Publishing Centre, Wellington

Macheix J-J, Fleuriet A, Billot J (1990) Fruit phenolics. CRC Press, Boca Raton

Manabe T (1982) Quantitative determination of total tannin in the astringent Saijyo persimmon before and after removal of astringency. Bull Hiroshima Agric Coll 7:79–85

Matsuo T, Ito S (1977a) On mechanisms of removing astringency in persimmon fruits by carbon dioxide treatment. I. Some properties of the two processes in the de-astringency. Plant Cell Physiol 18:17–25

Matsuo T, Ito S (1977b) Kaki tannin. J Chem Biol (Kagaku to Seibutsu) 15:732–736 (in Japanese)

Matsuo T, Ito S (1978) The chemical structure of kaki-tannin from immature fruit of the persimmon (*Diospyros kaki* L.). Agric Biol Chem 42:1637–1643

Nakabayashi T (1971) Studies on tannins of fruits and vegetables. VII. Difference of the components of tannin between the astringent and non-astringent persimmon fruits. J Jpn Soc Food Sci Technol 18:33–37

Piretti MV, Pistore R, Razzoboni C (1985) On the chemical constitution of kaki tannin. Ann Chim 75:137–144

Sugiura A, Harada H. Tomana T (1975) Studies on the removability of astringency in Japanese persimmon fruits. I. 'On-tree removal' of astringency by ethanol treatment. J Jpn Soc Hortic Sci 44:265–272 (in Japanese with English Summary)

Sugiura A, Harada H, Tomana T (1977) Studies on the removability of astringency in Japanese persimmon fruits. II. 'On-tree removal' of astringency by ethanol treatment. J Jpn Soc Hortic Sci 46:303–309 (in Japanese with English Summary)

Swain T, Hillis WE (1959) The phenolic constituents of *Prunus domestica*. I. The quantitative analysis of phenolic constituents. J Sci Food Agric 10:63–68

Taira S, Shimomura M, Oba S, Watanabe S (1990) Changes in soluble and total tannin content during fruit development and their relation to removal of astringency in Japanese persimmon (*Diospyros kaki* Thunb.). Abstr XXIII Int Horticultural Congr, 27 Aug–1 Sept 1990, Firenze, vol 2, no 3279

Van Buren J (1970) Fruit phenolics. In: Hulme Ac (ed) The biochemistry of fruits and their products, vol 1. Academic Press, London, pp 269–304

Yonemori K, Matsushima J (1984) Chemical characteristics of tannins from non-astringent and astringent type fruits of Japanese persimmon (*Diospyros kaki*) with particular reference to ultracentrifugal behavior. J Jpn Soc Hortic Sci 53:121–126 (in Japanese with English Summary)
Yonemori K, Matsushima J, Sugiura A (1983) Differences in tannins of non-astringent and astringent type fruits of Japanese persimmon (*Diospyros kaki* Thunb.). J Jpn Soc Hortic Sci 52:135–144 (in Japanese with English Summary)

Analysis of Fresh and Dried Apricot

J.M. WITHERSPOON and J.F. JACKSON

1 Introduction

The apricot (*Prunus armeniaca* L.) is one of the most desirable of the temperate tree fruits. It is delectable and versatile, being important for fresh production as well as for the canning, dehydrated and juicing industries (Burgos and Ramming 1993). Exceptional fruit quality consists of a balance of sugar and acidity as well as a strong apricot aroma (Mehlenbacher et al. 1983). Thus it is important to be able to accurately and reproducibly measure fruit quality characteristics.

The apricot is an important crop worldwide with total world production reaching 2.2 million tonnes in 1991. Turkey is the single largest producer of apricots with total production reaching approximately 430 000 tonnes in 1991. The USA produces 83 000 tonnes of apricots primarily for the fresh market, and Australia produces a total of 34 000 tonnes per year (FAO 1992).

The Austrlian dried apricot is world renowned for its quality and colour. Australia is the second largest producer of dried apricots in the world, producing 2750 tonnes per year; however, this makes up less than 10% of world production. Turkey is the largest producer of dried apricots producing 35 000 tonnes in 1989 (I. Thorn, pers. comm. 1993).

2 Fresh Apricot Assessment

2.1 Determination of Fruit Size

Size is an important criteria for market acceptance and thus accurate methods of determination are important in research. Fresh fruit size is typically expressed as a diameter measurement, determined by measuring the maximum diameter of the equatorial section of the apricot. Measurements of fruit size can be made on individual fruit using adjustable sizing rings, fruit sizing boards or callipers (Childers and Zutter 1977).

The most accurate size determination is made with sizing rings or callipers. When measuring size with callipers, the instrument is placed across the suture line to measure cheek diameter. The callipers are then adjusted so they just touch the surface of the apricot. To take a measurement using the sizing ring, it is placed

Modern Methods of Plant Analysis, Vol. 18
Fruit Analysis
Edited by H.-F. Linskens and J.F. Jackson

perpendicular to the suture at the point of maximum diameter; the ring is adjusted to touch the fruit surface and the reading taken. Alternatively, fruit can be passed through holes in a sizing board to determine the diameter of the fruit at its maximum point.

For the export market, fruit is sorted into three classes:

	Minimum size (mm)	Variation (mm)
Extra class	35	±5
Class 1	30	±10
Class 2	30	±10

(e.g. Australian Export Control Order 1991)

Fruit can also be graded on the basis of total fruit weight, stone included (Guerriero and Watkins 1984)

		(g)
1	Extremely small	<20
2	Very small	20–30
3	Small	31–40
4	Small/medium	41–45
5	Medium	46–55
6	Medium/large	56–60
7	Large	61–70
8	Very large	71–85
9	Extremely large	>85

2.2 Determination of Fruit Firmness

Fruit firmness is often used as a measurement for determining maturity. Penetrometers (Magness-Taylor and Effi-gi) measure the resistance of flesh to the penetration of a standard probe (plunger). The Magness-Taylor fruit pressure tester, also called the D.Ballauf pressure tester, is one of the more traditionally used manual testers. The Effe-gi (Italy) fruit tester is, however, becoming more popular because of its smaller size. The probes for both penetrometer types are of the same dimensions, but readings from each instrument differ (Abbott et al. 1976). Both the Magness-Taylor and the Effe-gi are supplied by the manufacturers with 11.1 mm and 7.9 mm diameter probes (plungers). The 7.9 mm probe is more appropriate for apricots because of the softness and size of the fruit; however, penetrometers fitted with an 11-mm probe have been used to determine firmness of apricots (Visagie 1985; Fourie et al. 1992).

The regression equation:

$$P_7 = P_{11}/1.16$$

can be used to relate penetrometer readings with the 7.9-mm probe (P_7) to those taken with an 11.1-mm probe (P_{11}) (Brown and Walker 1990).

Hand-held pressure testers (such as the Magness-Taylor or Effi-gi) are convenient to use because they can be carried into the field. Larger automated pressure testers, such as the Instron Universal Testing Instrument, can be used to measure fruit firmness if a higher degree of accuracy of required. The three instruments, Effe-gi, Magness-Taylor and the Instron Universal Testing Instrument do not give identical numerical values when used to determine the firmness of apricots. It is therefore necessary to specify the instrument when reporting pressure test values (Wills et al. 1981).

2.2.1 Procedure – Hand-Held Penotrometers

Penetrometer readings may be taken through the skin; however, removing the skin improves the accuracy (Bourne 1980). When determining firmness with a penetrometer fitted with a 7.9-mm probe, a 10–20 mm (diameter) disc of skin is removed with a sharp knife. The fruit is placed on a hard surface such as a desk or against a wall with the penetrometer held at a right angle to the fruit surface. Pressure is applied evenly until the probe reaches the scribed 7.9 mm line. Erratic motion should be avoided, because these are maximum force instruments and a momentary high force will be recorded as the test force (Bourne 1980). Mounting the instrument in a drill stand can reduce measurement variability because the penetration force is applied at a more consistent rate (Abbott et al. 1976; Kader 1985). The test can be carried out on either one or both cheeks, depending on accuracy required and evenness of ripening. If two or more readings are taken on any one piece of fruit, the mean value should be calculated (Abbott et al. 1976).

Readings of force rather than pressure are measured and should be reported in force units as newtons ($N = m\ kg\ s^{-2}$).

Conversion factors: pound-force = 0.2248 N
kilogram-force = 0.1002 N (Soule 1985).

Probe size should be kept constant for any series of tests, and all determinations for a given experiment should be made by one person to minimize variability.

2.2.2 Procedure – Instron Universal Testing Instrument

A 10–20-mm disc of skin is removed, the fruit is then mounted in a bevelled ring on a load cell. The Instron is fitted with a 7.9 mm diameter probe and operated at a penetration rate of 25.4 mm/min (Abbott et al. 1976). The probe penetrates the exposed flesh to the scribed depth of 7.9 mm, and punctures are made around the equatorial circumference of the fruit at two locations (Roberston et al. 1984). The measurement is expressed in kilograms, pounds or newtons.

2.2.3 Non-Destructive Firmness Test

Penetrometers are frequently used to quantify the firmness of apricots. However, pressure testing is not an ideal way of assessing the firmness of apricot flesh, because there is often a gap (air space cavity) between the stone and the flesh. When pressure is applied to the fruit, rather than measuring the firmness of the flesh, the flesh moves slightly so as to fill the cavity between the flesh and stone, resulting in readings that are not reproducible (F. Gathercole, pers. comm. 1993). All the penetrometer methods are destructive and therefore cannot be used in industry to grade fruit in terms of firmness. Furthermore, destructive measurements of fruit firmness do not allow selection of constant maturity and firmness for postharvest trial work.

A new non-destructive method which measures firmness based on impact response analysis is currently being tested in New Zealand. Fruit is dropped onto a load cell from a constant height normally between 10 and 20 mm above the cell. The bounce signal is then calibrated to a reading of firmness on a force/time characteristic. If required, these readings can be correlated to penetrometer readings. This process does not significantly damage the fruit, allowing its use in industry to routinely determine firmness (Biggs 1993).

2.3 Determination of Total Soluble Solids (TSS)

Total soluble solids (TSS) is a measurement of the sugar, organic acid and other soluble components in the juice of the apricot. TSS is useful as an indicator of maturity in conjunction with firmness. It is measured in terms of percent pure sucrose because sucrose is the major constituent of apricot fruit juice. TSS can be determined using a Brix hydrometer (measurement of specific gravity), or an Abbe or hand refractometer (measurement of the refractive index) (Childers and Zutter 1983).

If a bulk sample of juice is to be determined, a hydrometer is recommended. For smaller samples, the larger Abbe can be used for precise determination in a laboratory or the hand model can be used in the field. All instruments give readings in terms of °Brix calibrated as percent sucrose at 20 °C (Soule 1985). It must be remembered that there are other soluble solids which may alter the refraction of light, present in the juice (Byrne et al. 1991). However, because sucrose is the major constituent of the total soluble solids, readings give a reasonable reflection of sweetness (Kramer 1973).

Total soluble solid content of apricots ranges from 5–30 °Brix depending on the variety and maturity of the fruit (Mehlenbacher et al. 1983).

2.3.1 Sample Preparation for Tests

Randomly select ten apricots per tree. Remove the stone from flesh and macerate flesh with a blender. The pureed flesh is then sieved and the juice is collected. The juice is then used to determine soluble solid content and sugar content.

2.3.2 Procedure – Hydrometer

If large amounts of juice are available, the use of a calibrated hydrometer provides accurate readings quickly and inexpensively. Calibrate the hydrometer in terms of percentage sucrose using watery sucrose standard solutions at 20 °C. Then fill a graduated cylinder with 200 ml of juice at 20 °C, insert the hydrometer and read the value as °Brix, which is equivalent to percentage of soluble solids (Kramer 1973).

2.3.3 Procedure – Refractometer

Hand refractometers are extensively used in industry and research, because they are accurate, quick and easy to use. Readings are made by placing one drop of juice (by squeezing a slice of fruit or using prepared juice extract) onto the hinged plate of the instrument. Face towards the light and look through the eye piece of the instrument and the value is record as °Brix (Childers and Zutter 1977). It is essential to clean the refractometer with distilled water after each reading to ensure that no particles remain on the hinged plate (Kader and Mitchell 1989).

The temperature of the juice is important in determining the accuracy of the reading. If the juice is heated, the particles in the solution expand, resulting in a lower reading, or if the solution is cooled, the solution becomes denser and the reading is higher. Some refractometers have a temperature compensation capability or a thermometer attachment so they can be adjusted for temperature variations.

2.4 Sugar

Sucrose is the major sugar in apricots, followed by glucose then fructose, while other sugars such as maltose are present only in minute amounts (Wills et al. 1983). Therefore methods for determination of glucose, fructose and sucrose only are given.

Typical sugar values (per 100 gram edible portion) found in apricots are:

Glucose	Fructose	Sucros
2.1 g	0.5 g	5.1 g

(Wills et al. 1987)
A kit to carry out the sucrose / glucose / fructose test based on the principal outlined above, is available from Boehringer Mannheim. Cat. No. 716-260.

2.4.1 Determination of Glucose and Fructose

This procedure, based on that described by Southgate (1991), can be used for the measurement of glucose only or to determine the individual amount of glucose and/or fructose in a mixture. Proteins interfere in both reducing and colorimetric sugar methods so it is desirable to remove the bulk of protein before attempting analysis. Deproteinize the solution with ice-cold 1 N perchloric acid in the proportion of 1:1, centrifuge and immediately neutralize the supernatant with 2 M KOH. Leave for 20 min in an ice-bath and filter off the precipitated potassium perchlorate. The solution should contain 0.4–0.8 g glucose + sucrose per litre; stronger solutions must be diluted. Remove excessive concentrations of glucose by treatment with glucose oxidase (Southgate 1991).

Reagents

Triethanolamine buffer (Tra buffer).
Dissolve 11.2 g triethanolamine. HCI and 0.2 g $MgSO_4 \cdot 7H_2O$ in 150 ml of distilled water, adjust to pH 7.6 with approximately 4 ml 5 M NaOH, then make up to 200 ml with distilled water. Bring buffer to 20–25 °C before use. (The buffer is stable for 4 weeks when stored at +4 °C).

Nicotinamade-adenine dinucleotide phosphate solution (NADP).
Dissolve 50 mg NADP-Na_2 in 5 ml of distilled water. (Stable for 4 weeks when stored at +4 °C).

Adenosine-5′-triphosphate solution (ATP).
Dissolve 250 mg ATP-Na_2H_2 and 250 mg $NaHCO_3$ in 5 ml of distilled water. (Stable for 4 weeks when stored at +4 °C).

Hexokinase/glucose-6-phosphate dehydrogenase (HK/G6P-DH).
Mix 0.5 ml of hexokinase solution (2 mg/ml) with 0.5 ml of glucose-6-phosphodehydrogenase solution (1 mg/ml). (Stable for 1 year when stored at +4 °C)

Phosphoglucose isomerase (PGI).
Suspension is used undiluted 2 mg/ml. (Stable for 1 year when stored at +4 °C)

Procedure

Zero spectrophotomenter against distilled water, check zero at frequent intervals throughout experiment. Pipette 3 ml of triethanolamine buffer at 20–25 °C into a cuvette (1 cm path length) and add 0.10 ml of NADP solution, followed by 0.10 ml of ATP solution and 0.20 ml of sample. Mix the contents by inversion and measure the absorbance (optical density) at 340 nm (E_1). Add 0.02 ml of HK/G6P-DH suspension and, after mixing, allow the reaction to go to completion for 10–15 min, then measure the optical density again (E_2). If the reaction has not reached completion after 15 min, continue to read the absorbance at 2-min intervals until absorbance is stable over the 2-min period. Add 0.02 ml PGI solution and mix, then read the absorbance after 10 min (E_3).

Determine the absorbance difference:

$E_2 - E_1$ = E glucose (density change due to glucose)
$E_3 - E_2$ = E fructose (density change due to fructose).

Calculation

The volume of the mixture for measuring glucose is 3.42 ml and for fructose 3.44 ml, with a 0.2 ml sample:

E glucose × 0.495 F = x (g glucose/l)
E fructose × 0.495 F = y (g fructose/l),

where F is the dilution factor.

If the spectrometer is not reliable at 340 nm, it often can be used at the higher wavelength of 366 nm. If this is the case and all readings are taken at 366 nm, the following calculations are used:

E glucose × 0.933 F = x (g glucose/l)
E fructose × 0.938 F = y (g fructose/l),

where F is the dilution factor of the sample.

2.4.2 Determination of Sucrose by Invertase Hydrolase

The protein can be removed by the addition of 1 N perchloric acid in the proportion of 1:1 as explained in Section 2.4.1. The solution should contain 0.4–0.8 g glucose + sucrose per litre; stronger solutions must be diluted. Remove excessive concentrations of glucose by treatment with glucose oxidase (Southgate 1991).

Reagents

Acetate buffer.
Dilute 6.7 ml 1 M NaOH and 13.5 ml 1 M acetic acid with 180 ml of water. The pH should be 4.6. Bring buffer to 20–25 °C before use. (Buffer is stable for 1 year when stored at +4 °C)

Beta-fructosidase solution.
Dissolve 10 mg of beta-fructosidase lyophilisate in 2 ml of distilled water. (Stable for 1 week when stored at +4 °C)

The following solution are the same as those used in Section 2.4.1.

Triethanolamine buffer (Tra buffer).
Bring buffer to 20–25 °C before use. (The buffer is stable for 4 weeks when stored at +4 °C).

Nicotinamade-adenine dinucleotide phosphate solution (NADP).
(Stable for 4 weeks when stored at +4 °C)

Adenosine-5′-triphosphate solution (ATP).
(Stable for 4 weeks when stored at +4 °C)

Hexokinase/glucose-6-phosphate dehydrogenase (HK/G6P-DH).
(Stable for 1 year when stored at +4 °C)

Phosphoglucose isomerase (PGI).
(Stable for 1 year when stored at +4 °C)

Procedure

Mix 0.20 ml of sample with 0.50 ml of acetate buffer and 0.002 ml of the beta-fructosidase solution in a cuvette(1 cm path-length) and leave it to stand for 15 min at 20–25 °C. Then add 2.50 ml of Tra buffer, followed by 0.10 ml of the NADP solution and 0.10 ml of the ATP solution. After mixing, measure the absorbance at 340 nm (E_1). Then add the 0.02 ml HK/G6P-DH solution and leave the mixture for 10–15 min before measuring the absorbance again (E_2). If the reaction is not complete at this time, measure the absorbance at 2-min intervals until the increase per interval is constant. Extrapolate E_2 to the time of addition of the hexokinase.

Calculation

$$[(\text{Sucrose} + \text{glucose})\text{-free glucose}] \times 2 \times 0.95 = Z\,\text{g sucrose/l.}$$

With a total volume of 3.44 ml for the sucrose reaction and 3.42 ml for the glucose reaction, and a sample volume of 0.20 ml the following calculation applies:

$$[(E2 - E1_{(\text{sucrose test})} \times 0.497) - (E2 - E1_{(\text{glucose test 2.4.1})} \times 0.495)] \times 2 \times 0.95 = Z\,\text{g sucrose/l.}$$

Alternatively, values can be measured using a wave length of 366 nm in which case the following calculations are used:

$$[(E2 - E1_{(\text{sucrose test})} \times 0.939) - (E2 - E1_{(\text{glucose test 2.6.1})} \times 0.933] \times 2 \times 0.95 = Z\,\text{g sucrose/l.}$$

(Boehringer Mannheim 1989; Southgate 1991)

2.5 Acidity

The two methods frequently used to measure the acidity of apricots are percentage of total titratable acidity, usually in terms of malic or citric acid, and hydrogen ion concentration or pH. The titratable acidity and PH are not directly related, as the latter depends on the concentration of free hydrogen ions and buffering capacity of the extracted juice (Wills et al. 1981). Titratable acidity measures the total available amounts of hydrogen ions in the solution (Kader 1985).

2.5.1 Determination of Total Titratable Acidity

Weigh 10–20 g of macerated sample (Sect. 2.3.1) into a tared 250-ml flask and record the exact weight of the sample. Add distilled water to the flask to bring the volume to 100 ml. Colorimetric titration may be used if the sample is not highly coloured or electrometric titration for highly coloured samples.

Colorimetric Titration

Fill a burette (50 ml graduated in 0.1 ml) with 0.1 N sodium hydroxide then add three to five drops of phenolphthalein indicator to the sample. While stirring or swirling the titration flask carefully, add alkali from the burette until the colour changes from colourless to a faint pink throughout the solution. The colour should remain after mixing for 30 s.

Electrometric Titration

Place the electrodes of pH meter in the sample, if the electrode is not covered completely add distilled water. Add 0.1 N sodium hydroxide from the burette while stirring the flask. Continue to add the NaOH until the pH reaches 8.3 on the pH meter.

Calculation of Results

For a 10-g sample multiply the number of millilitres of 0.1 N NaOH used in the titration by the appropriate factor for the acid listed below to obtain the percentage of total titratable acid. For a 20-g sample divide the millilitres of 0.1 N NaOH by 2 before multiplying by the acid factor.

Acid NaOH	Equivalent	Weight Factor for 0.1 N
Acetic	60.05	0.06005
Lactic	90.08	0.09008
Malic[a]	134.09	0.06705
Citric (anhydrous) (Kramer 1973)	192.12	0.06404

[a] Malic or citric acid is often used to determine total titratable acid in apricots. The acid concentration is expressed as milliequivalents of malic acid per millilitre of juice (mEq H+ /ml) (Visagie 1985; Byrne et al. 1991) .

A range of from 0.3–2.6 malic acid/100 ml juice with an average of 1.6 is normally found for apricot (Mehlenbacher et al. 1983; Visagie 1985).

2.5.2 pH

Strain 10–15 ml of the pulp mixture (Sect. 2.3.1) into a pH testing cup. Calibrate the pH meter according to the instruction supplied with the meter. Rinse the electrode with distilled water, then place the electrode into the sample, ensuring that the tip of the electrode is completely covered (if it is not, add distilled water to the sample until the electrode is covered). Record the pH value, repeat the test and average the readings. Generally, the pH for Australian apricots is approximately 4.1 (F. Gathercole, pers. comm. 1992).

2.6 Determination of Juice Viscosity

Fruit is cut in half and stone removed, the flesh is pulped using a fruit blender, juice is collected by filtration through a Whatman No. 541 filter paper. A 0.2-ml sample is drawn up a 0.2-ml graduated pipette that has been previously calibrated against Brookfield viscosity standards. The time taken for 0.1 ml of the juice to flow out of the pipette under gravity is recorded with a stop watch with accuracy to 0.01 s. The mean of four reading should be recorded. Ensure that the sample is thoroughly mixed prior to each measurement. Check calibration at frequent intervals. Calculate the viscosity according to the calibration of the pipette. Juice viscosity is measured in centipoise with a range of 7–13 usually recorded for apricots (Brown and Walker 1990).

2.7 Colour

2.7.1 Colour Comparators

The colour of apricots can be determined by comparison with visual standard colour plates such as the Royal Horticultural colour charts. A disadvantage of colour plates is that they are two-dimensional and the eye sees colour in three dimensions. hue, value and chroma.

Furthermore, the use of colour plates is limited due to errors by the observer and interpretation from one observer to another (Moore and Janick 1983).

When measuring colour with colour comparators, three colour parameters have to be determined, ground or skin colour, flesh colour and the blush which may occur over the predominate ground colour. Flesh colour can be determined either by making a tangential cut with a sharp knife to expose an area of flesh, about 20 mm in diameter (the depth should be standardized for each sample), or by cutting the apricot in half along the suture, removing the stone and observing the internal colour (excluding the stone cavity as it is not a true indication of the flesh colour). These three colour parameters are then compared to the standard colour plates and fruit colour is then recorded (Bourne 1980; Soule 1985; Kader and Mitchell 1989; McGuire 1992).

2.7.2 Light Reflectance Meters

Reflectance meters such as tristimulus colorimeters measure colour on the basis of the amount of light reflected from the surface of the fruit (Kader 1985). Several types of tristimulus colorimeters are available that measure the colour of fruit depending on lightness, red to blue tone and yellow to green tones. Two common types are the Hunter Colorimeter system and the various Minolta Chroma Meters (CR-300 series), both types use the L, a, b coordinates (Brown and Walker 1990; McGuire 1992).

The L scale ranges from no reflection ($L = 0$) to perfect diffuse reflection ($L = 100$) and is a measure of light or darkness. The a scale ranges from negative values for green to positive values for red, and the b scale ranges from negative values for blue to positive values for yellow. The term used to distinguish the red, yellow, blue and green colours is called hue (Delwiche and Baumgardner 1983; Minolta 1991). The colour meters qualitatively identify colours and give it a numerical value without relying on an individual's colour perception, as colour plates do. The colour of an object will appear differently, depending on the angle from which the object is illuminated and the angle from which it is viewed. For the best result with a Minolta Chroma Meter (Minolta Model CR-300), the colour of fruit is determined using the d/o system (diffuse illumination/0° viewing geometry). It provides measurements of a wide surface which correlate well with the colour seen under diffuse light conditions. An area of 8 mm in diameter is measured. The instrument is calibrated with a white reference plate or reference colour suggested by supplier, and measurements are recorded using the L, a and b colour coordinates. Minolta a and b values can be used to compute values for hue angle $[\theta = \tan^{-1}(b/a)]$. This parameter is effective for visualizing the colour appearance of food products (Roberston et al. 1984). For a fresh fruit sample, four readings should be taken for each part of the fruit corresponding to ground, blush and flesh colour.

If the Hunter colour difference meter (Model D25A-2, F,STH,H, Hunter Associates Laboratory) is used, it employs 45° diametrically opposed illumination and a 6.4-mm viewing aperture. Due to the directional nature of the colorimeter illumination and small colour variation within the designated surface area, each fruit should be rotated 90° on axia, and a second within the same area should be averaged with the first (Delwiche and Baumgardner 1985).

Each colorimeter system uses different coordinates, yet all can be compared to the colour set X, Y, Z used by the International Commission on Illumination and Munsell systems (Moore and Janick 1983).

Note. The a colour coordinate (green to red) is the principal coordinate used for indicating fruit maturity (Brown and Walker 1990).

2.8 Determination of Pectic Acid

This procedure is based on that described by Southgate (1991) in which the pectic substances are saponified to remove ester groups, pectic acid is then precipitated and weighed.

Reagents

Sulphuric acid	0.5 m
Sodium hydroxide	10% (w/v)
Ethanol	5% (v/v)
Hydrochloric acid	10 ml plus 25 ml of water

Procedure

Whole fruit is cut in half, the stone is removed and the halves are finely chopped before extraction. Three hundred g of the chopped sample is mixed with 800 ml of water and heated to the temperature of boiling water for 1 h. Transfer the mixture to a 2-l flask and make up to volume. Place a 200-ml, filtered portion of the extract in a beaker (600 ml). and add 10 g of sucrose if the solution is low in sucrose. Add 200 ml of ethanol with constant stirring and allow the precipitate to settle. Filter off the precipitate on a qualitative paper and wash with ethanol.

Return the precipitate to its original beaker with hot water and evaporate to a volume of about 40 ml, then cool the solution to 25 °C. Redissolve any precipitate which forms at this stage by acidifying the solution and warming. Dilute 5 ml of 10% (w/v) NaOH to 50 ml and add to the solution. (The volume of the sodium hydroxide used depends on the bulk of the precipitate.)

Leave the mixute for 15 min. then add 40 ml of water followed by 10 ml dil. HCl and boil the mixture for 5 min Filter the precipitated pectic acids and wash with hot water.

Return the precipitate to the beaker and repeat the saponification. Wash the precipitate until the filtrate is only slightly acidic. Transfer the precipitate to a platinum dish and dry at 100 °C, weigh and then ignite and re-weigh. Take the loss in weight as pectic acid. A value of approximately 0.5 g pectin/100 g fresh weight is normally expected for apricots (Southgate 1991).

A colloidal precipitate that is difficult to filter is indicative of saponification that is incomplete and should be repeated. (AOAC 1990; Southgate 1991)

Note. Samples for pectic acid analysis require careful preservation to ensure that the pectin is not degraded or lost during sample handling and storage. Samples should be stored at −20 °C in sealed containers if they are not to be analyzed immediately, but some physical changes may occur that could interfere with the subsequent analysis. Analysis of fresh samples is undoubtedly the preferred approach.

2.9 Determination of Starch

As fruit matures, starch is converted into sugar. Decreasing amounts of starch may therefore be used as a relative indicator of maturity. The purple colour of starch when it is reacted with iodine (starch-iodine test) is used as an indicator of starch content. To prepare iodine solution dissolve 6 g potassium iodide in 400 ml of water, then add 1 g iodine (I_2).

Cut an apricot into sections then dip a slice of fruit from the blush and unblushed side of each fruit into the iodine solution for 1 min. Remove the slices from the solution and let stand for 2 min. Wash each slice quickly in water and determine the percentage of starch. Starch will show up as dark blue area, white areas will contain sugar (Childers and Zutter 1977).

Note. Fully ripe apricots contain no starch (Jones et al. 1990).

3 Dried Apricots

Quality control procedures that are unique for dried fruit products include: moisture content determination, sulphur dioxide determination (pre- and post-drying), physical characteristics of dried fruit and drying ratio (Woodroof and Luh 1975).

3.1 Size

Fruit can be graded into sizes using a standard grader system. If a large amount of dried fruit is to be sized/graded an automated sieve system can be used. Different sieves are used and each sieve has its own mesh size. Medium-sized apricots must pass through a sieve with a mesh of 25 mm but not 15 mm, large apricots must pass through 35 mm but not through 25 mm and extra large not through the 35 mm. Jumbo dried apricots must pass over a 40-mm sieve.

Sizing of dried apricots are based on:

Jumbo	90% of fruit must pass over 40 mm riddle
Extra large	90% of fruit must pass over 35 mm riddle
Large	90% of fruit must pass over 25 mm riddle
Medium	90% of fruit must pass over 15 mm riddle

(Australian Export Control Orders no. 8 schedule 21 989)

Callipers can be used if more accurate size measurements are required (see Sect. 2.1).

3.2 Determination of Drying Ratio

Drying ratio is determined by calculating the mass of a sample of fruit before and after drying (stone included in fresh weight only). It is calculated by dividing the fresh weight by the dry weight and expressed as a ratio. Moisture content of the dried fruit sample is critical to the calculation and is normally measured and adjusted to 13%. Where multiple drying ratios are required, dried fruit samples can be equilibrated to the desired moisture content in a controlled humidity temperature cabinet. Small variations in the moisture content of fruit will occur with differing fruit quality (Dahlenburg, unpubl. 1987). Total soluble solid content is negatively correlated with drying ratio, and the drying ratio can be predicted from the Brix value of fresh fruit as determined in Section 2.3 (Fourie et al. 1992). The average drying ratio for Australian dried apricots is 6:1 i.e. 6 tonnes of fresh fruit will dry to approximately 1 tonne of dried fruit, the lower the drying ratio the better the quality the fruit (F. Gathercole, pers. comm. 1993).

3.3 Determination of Moisture

The moisture content of fresh apricots averages 85.5% per 100 g fruit with a range of 91.5 to 81.5%, while the moisture content for dried apricots ranges from 10 to 25%. The moisture content of dried apricots is of paramount importance because of the preservation effect the lowering of moisture has on the fruit. Several methods are available for determining the moisture in dried apricots. The vacuum oven method involves the measurement of weight loss due to the evaporation of water and takes up to 8 h to gain a result. A more rapid method for calculation of moisture in dry products can be made by using an infra-red moisture meter or a Calipco meter which measures electrical resistance. the Calipco meter was developed by C.D Fisher in 1933 for the California Dried Fruit Association (Joslyn 1970).

3.3.1 Vacuum Oven

This procedure is based on that described in the 15 th edition of AOAC (1990) for moisture in dried fruit.

Take approximately 50 g dried fruit and mince twice with a mincer. With fruit rich in sugar such as apricots, use approximately 5 g of sample and dry in a dish with 2 g of finely divided asbestos. Dried sand, which has previously been washed in hydrochloric acid and then rinsed thoroughly with distilled water, may be used in place of asbestos. Moisten the sample and asbestos thoroughly with 3–5 ml of hot water, then mix sample and asbestos together with a spatula. Wash the spatula with hot water to remove the sample residues, let the residues and the water fall into the sample dish. Heat the open dish in a boiling water bath or steam bath to evaporate the moisture to dryness.

Ensure that the prepared sample is spread as evenly as possible over the bottom of the non-corrosive metal sample dish (8.5 cm diameter), which has a tight fitting lid. Weigh the dish + sample + lid and dry sample in oven for 6 h at 70 ± 1 °C under constant pressure 100 mm Hg (13.3 kPa), the oven should not be opened during this time (note: the lid is also placed in the oven; however, do not cover the dish). The metal dish must be in direct contact with metal shelf of the oven. During drying admit a slow current of air (about two bubbles/s) to the oven, the air having been dried by passing it through H_2SO_4.

After drying, remove dish and cover it immediately with its lid and place in a desiccator to cool. Afte cooling to 20–25 °C, weigh the covered dish to the nearest 0.01 g. The sample and lid should be returned to the oven for a further 2 h and rewieghed to ensure that a constant weight is attained (Wills et al. 1980).

Calculation of the water content.

The moisture content of the sample as a percentage of the total mass is calculated as follows:

$$\text{moisture content} = \frac{(M_1 - M_2)}{(M_1 - M_0)} \times 100,$$

where M_0 is the mass in g of the empty dish with its lid and the asbestos it contains; M_1 is the mass in g of the dish with its lid, the asbestos and the sample before drying; M_2 is the mass in g of the dish with its lid and the test sample after drying.

Duplication determinations should agree to 0.2% moisture content.

3.3.2 Infra-Red Moisture Meter

The sample must be finely ground and passed through a 10-mesh screen. A given weight is dried under an infra-red lamp, and the loss of weight is recorded by a sensitive scale on the tester. This test provides results in about 7–10 min. However, the results will not be as accurate as with the vacuum oven method (Woodroof and Luh 1975).

3.4 Colour

Colour of fruit can be determined by three methods

Colour extraction
Colour comparators (Sect. 2.7)
Light reflectance meters (Sect. 2.7).

3.4.1 Apricot Colour Extraction

The colour extraction method measures darkness of the fruit via optical density rather than colour of the whole fruit as such.

Weigh out 15.00 g (±0.01) of fruit. use either minced fruit or thinly cut slices (2 mm width) from across the centre section of the dried fruit. Add 200 ml of 50% ethanol, place on a shaker for 15 h. Filter sample through a Whatman No. 40 filter paper into vials. Blank the spectrophotometer on 50% ethanol, check zero regularly. Read absorbance with a spectrophotometer at 440 nm wavelength (McBean and Wallace 1967). A reading of >0.3 is recorded as unacceptably dark in the dried fruit industry (A. Dahlenberg, pers. comm. 1993).

3.5 Sulphur Dioxide

Sulphur dioxide has been used in the preservation of food products for many centuries (Barnett 1985). It is routinely used for drying apricots because it is inexpensive and effective. Sulphur in the form of free sulphur dioxide is mostly responsible for the preservative action in fruit. There are four main reasons for adding sulphur to preserve fruit:

1. Sulphur dioxide is an effective inhibitor of the non-enzymatic browning that occurs during storage of dried apricots.
2. The enzymatic browning of fruit if prevented.
3. The micro-organisms are inhibited by the presence of sulphur dioxide.
4. It has antioxidant properties that are useful in the preservation of the carotenoids present (Barnett 1985).

Residual sulphur may not exceed 2000 ppm for the world market (World Health Organization). A level of 2000 ppm of sulphur dioxide indicates that the dried product should maintain acceptability for up to a year in storage at ambient temperature and should remain of good quality. Darkening occurs rapidly in storage if the sulphur level in the dried product is lower than 1500 ppm (McBean et al. 1967).

3.5.1 Sulphur Content of Freshly Sulphured Apricots

Determined through direct titration with starch indicator and iodine.

This method is based on that described by Dahlenburg et al. (1986).

10% Sodium hydroxide. Dissolve 10 g of NaOH in 100 ml of water (caution)

20% Sulphuric acid. Dissolve 20 g of H_2SO_4 in 10 ml of water (caution)

0.02 N Iodine. Dilute 0.1 N Iodine prepared from Volucon ampoules by accurately transferring 200 ml of 0.1 N into a 1-l volumetric flask and make up to the mark with distilled water.

Vitex or soluble starch powder.

Procedure

Prepare a 400-ml beaker of known mass with 190 ml of distilled water and 10 ml of 10% NaOH. Add six freshly sulphured apricot halves to the jar ensuring they are covered by the liquid, weigh the jar to determine total mass of apricot halves and record mass.

(Note. Fruit samples can be kept for up to 3 days in a well-sealed jar, without a significant loss of sulphur dioxide; however, testing should be conducted as soon as possible.)

Macerate the fruit for 2 min using a macerating type blender (vitamises are not suitable). Transfer 10 ml of the pureed fruit into a 500-ml sealable jar (ensuring all is transferred), then add 100 ml of distilled water and 10 ml of 10% NaOH. Seal jar and shake for 1 min, then allow to stand for 20 min (this time is critical). Add 20 ml of 20% H_2SO_4 and 1 g of Vitex to the jar and titrate using 0.02 N iodine to end point of navy blue colour. The end point can be confirmed by sealing the jar and shaking for 15 s; the blue colour should remain.

Calculation

$$\text{Sulphur dioxide (ppm)} = 64 \times \text{I} \times \left(1 + \frac{200}{\text{fruit weight (g)}}\right)$$

where I is the volume (ml) of 0.02 N iodine solution added.

3.5.2 Sulphur Content of Apricots after Drying

Direct titration method for sulphur dioxide with Vitex/iodine

This method is based on that described by McBean (1985). Mince 500 g of dried apricots, place an accurate sub-sample of 50 g fruit in a glass 500-ml sealable jar, record the mass of sample + jar to an accuracy of 0.01 g. Add 10 ml of 10% NaOH and make mass up to 250 g using distilled water (tap jar to remove any displaced air), record the mass. Macerate mixture for 2 min or until the puree is of a thick creamy consistency without any lumps. Hard or unprocessed fruit may be left to stand for up to 24 h or overnight to soften fruit. If left to stand, puree the mixture for a further 2 min or until finely pureed, prior to use. Place 50 g of pureed sample in a 500-ml glass jar, rinsing measuring cylinder with 100 ml of distilled water to collect all puree. Add 10 ml of 10% NaOH to jar, seal and shake vigorously, then allow to stand for 20 min (this time is critical). Add 1 gram of Vitex or other starch indicator and 20 ml of 20% H_2SO_4. Titrate immediately with 0.02 N iodine until the indicator changes to navy blue. The end point can be confirmed by sealing the jar and shaking for 15 s; the blue colour should remain.

Calculation

$$\text{Sulphur dioxide (ppm)} = \text{I} \times 64,$$

where I is the volume (ml) of 0.02 N iodine solution added.

3.5.3 Modified Monier-William SO_2 Method

The official (AOAC 1990) method for measuring sulphur dioxide is the Monier-Williams technique. The method is lengthy, requiring 3 h of refluxing in acid solution, and is suitable when accuracy of sulphur dioxide content is essential. The method described here is a shortened version of the Monier-Williams method developed by Dahlenburg (1984). This method recovers greater than 95% of the sulphur dioxide released under the longer AOAC (1990) method.

The basic modification made to the Monier-Williams technique was to reduce the refluxing time from 3 h to 50 min. With this modification, a dried fruit sample can be analyzed in approximately 70 min. There is also a slight modification to the apparatus (Fig. 1).

Reagents

0.25% Methyl Red
Dissolve 0.25 g of methyl red in 100 ml of ethanol.

0.1 N Sodium hydroxide
Dissolve 4.2 g A.R. NaOH in 1 l of distilled water.

Hydrochloric acid
Mix gently with continual stirring 100 ml of concentrated HCL with 200 ml of distilled water.

3% Hydrogen peroxide
Dilute 100 vol H_2O_2 (30%) to 6% by mixing 200 ml of 100 vol H_2O_2 with 500 ml of distilled water. Test with known normality $KMnO_2$ and calculate H_2O_2 concentration as shown below:

$$1\,\text{ml}\ 1\,\text{N}\ KMnO_4 = 17.01\,\text{mg}\ H_2O_2;\ \text{diluted to give}\ 3\%\ H_2O_2.$$

Procedure

Assemble apparatus as shown in Fig. 1, lightly greasing all joints with petroleum jelly. Remove gas inlet tube and adaptor and place 20 g of minced dried fruit in flask. Add 400 ml of distilled water and reinsert gas inlet, being sure not to block the end with the minced dried fruit. Half fill the gas absorption vessel with 20–25 g of glass beads and add 10 ml of 3% H_2O_2. Place 90 ml of hydrochloric acid into separating funnel and add to the flask with system fully closed. If necessary, force the acid solution into the system by connecting a second line attached to a dry N_2 cylinder or by blowing air into a plastic tube on an adaptor placed on top of the funnel.

Turn off cooling water to the condenser. Turn on N_2 gas supply to give a steady stream of bubbles in the flask (three bubbles/s). Peroxide should also commence bubbling; if not, check for leaks in joints.

Turn on heating mantle and reflux for 50 min after boiling and refluxing has commenced. At completion of refluxing, first remove gas absorption adaptor and transfer to 250 ml beaker all H_2O_2 and beads. Be sure all is removed from the adaptor. Turn off heat and after cooling, turn off condenser and water and clean for next sample. Add three drops of methyl red to beaker containing H_2O_2 and titrate against 0.1 N NaOH to colour change (clear yellow colour).

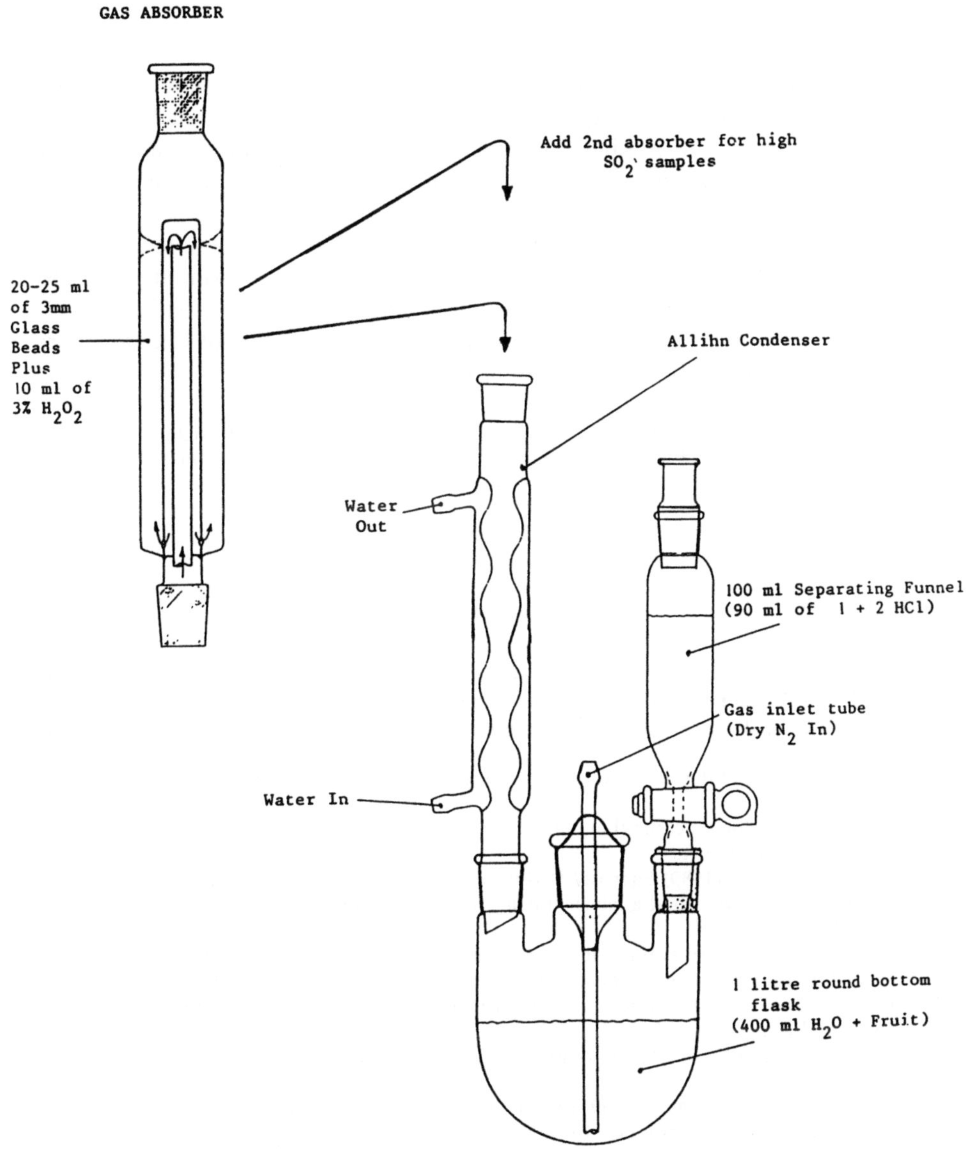

Fig. 1. Modified Monier-Williams apparatus as used by the Department of Primary Industries, South Australia

Calculation

$$SO_2 \text{ content (ppm)} = (1000 \times C \times T)/WT,$$

where C = 3.203; mg SO_2/ml of 0.1 N NaOH, T = titre (ml) of 0.1 N NaOH, and WT = mass (grams) of fruit sample.

3.6 Storage Life

Dried fruit must be capable of being stored for up to 12 months to allow continuity of supply to the market. During this time, fruit can deteriorate in quality, so it is important to be able to predict the storage life of a dried apricot sample. The storage life prediction test is based on the condition in which fruit would be after 4 months storage at 25 °C and 65% relative humidity.

Place 100 g of dried apricots in an open net bag. Then put sample in oven at 45 °C and 70% relative humidity for 7 days. The fruit is removed from the oven and the colour is detemined using the method described in Section 3.5 (Rettke 1993). If the reading is greater than 0.3, it is considered that the fruit will not store for more than 4 months (Dahlenberg, pers. comm. 1993).

Acknowledgements. The authors wish to thank Mr. M. Rettke of the South Australian Department of Primary Industries for advice and encouragement during the preparation of this chapter, and the Dried Fruits Research and Develpment Council for financial support.

References

Abbott JA, Watada AE, Massie DR (1976) Effi-gi, Magness-Taylor, and Instron Fruit Pressure Testing devices for apples, peaches, and nectarines. J Am Soc Hortic Sci 101:698–700

AOAC (1990) Official methods of analysis of the association of official analytical chemists, 15th edn. Association of Official Analytical Chemists, Arlington, VA

Barnett D (1985) Sulphites in foods: their Chemistry and analyses. Food Technol Australia 37:503–505

Biggs T (ed) (1993) The machine that checks the firmness of fruit. In: Good fruit and vegetable. Rural Press, Melbourne

Boehringer Mannheim (1989) Sucrose/ D-Glucose/ D-Fructose, Cat no 716 260. In: Methods of biochemical analysis food analysis. Boehringer Mannheim, Biochemica, W-Geramany, pp 134–139

Bourne MC (1980) Texture evaluation of horticultural crops. Proc symp Sensory and objective evaluation of horticultural crops. HortScience 15:51–57

Brown GS, Walker TD (1990) Indicators of maturity in apricots using biplot multivariate analysis, J Sci Food Agric 53:321–331

Burgos L, Ramming DW (1993) Apricot production and cultivar situation in California. Fruit Var J 47:91–94

Byrne DH, Nikolic AN, Burns EE (1991) Variability in sugars, acids, firmness, and colour characteristics of 12 peach genotypes. J Am Hortic Sci 116:1004–1006

Childers NF, Zutter HA (1977) Modern fruit science: laboratory manual. Somerset Press, New Jersey, pp 174–177

Dahlenburg AP (1984) Sulphur dioxide measurement techniques for dried fruits. Department of Agriculture, Adelaide. Rep 11/84

Dahlenburg A, Glenn T, Hill J (1986) Sulphur dioxide analysis methods for dried fruits. Department of Agriculture, Adelaide. Rep E/4/86

Delwiche MJ, Baumgardner RA (1983) Ground color measurements of peach. J Am Soc Hortic Sci 108:1012–1016

Delwiche MJ, Baumgardner RA (1985) Ground colour as a peach maturity index. J Am Soc Hortic Sci 110:53–57

FAO (1992) FAO Statistic series no 104 Production. Food and Agriculture Organization of the United Nations, Rome, vol 45, pp 164

Fourie PC, Hansmann C, Wium GL (1992) Effects of fresh fruit characteristics and cold storage on the quality of dried apricots and peaches. J Hortic Sci 67:59–65

Guerriero R, Watkins R (eds) (1984) Revised descriptor list for apricots (*Prunus armeniaca*). IBPGR, Rome, 23 pp

Jones GP, Briggs DR, Wahlquist ML, Flentje LM, Shiell BJ (1990) Dietary fibre content of Australian foods. 3. Fruits and fruit products. Food Aust 42:143–145

Joslyn M (1970) Moisture content and total solids. In: Stewart KK, Whitaker JR (eds) Mordern methods of food analysis. AVI Publishing, Connecticut

Kader AA (1985) Quality factors: Definitions and evaluation for fresh horticultural crops. In: Kader AA, Kadmire RF, Mitchell KG, Reid MS, Sommer NF, Thompson JF (eds) Postharvest technology of horticultural crops. Agriculture & Natural Resources Publications, Division of Agriculture and Natural Resources, University of California, Oakland, pp 118–121

Kader AA, Mitchell FG (1989) Maturity and Quality. In: LaRue JH, Johnson RS (eds) Peaches, plums, and nectarines. Growing and handling for fresh market. Oaklands, California

Kramer A (1973) Fruits and vegetables. In: Kramer A and Twigg BA (eds) Quality control for the food industry. AVI Publishing, Connecticut, pp 170–174

McBean D McG (1985) Drying and processing tree fruits. Department of Agriculture South Australia Government Printer, Australia, Bulletin 1/85, pp 23–24

McBean D McG, Wallace JJ (1967) Stability of moist-pack apricots in storage. CSIRO Food Preserv 7:29–35

McGuire RG (1992) Reporting of objective color measurements. Hortscience 27:224–1225

Mehlenbacher SA, Cociu V, Hough LF (1983) Apricots (Prunus). In: Moore JN, Ballington JR Jr (eds) Genetic resources of temperate fruit and nut crops. International Society for Horticultural Science, Wageningen, pp 65–66

Minolta Camera Co Ltd (1991) Chroma meters. Japan

Moore JN, Janick J (1983) Methods in fruit breeding. Purdue University Press, West Lafayette, IN

Rettke M (1993) Prediction the storage life of dried apricots. ADFA News, Australia

Roberston GL, Koopmanschap EA, Scrivens CA (1984) Comparison of instruments and sensory panel methods for measuring kiwifruit firmness. J Texture Stud 15:275–283

Sould J (1985) Glossary for horticultural crops. John Wiley, New York, pp 570–582

Southgate DAT (1991) Determination of food carbohydrates. Elsevier Applied Science, London, pp 144–148

Visagie TR (1985) Optimum picking for apricots – preliminary results and recommendations. The Deciduous Fruit Growers, Capetown, vol 45, No 11, pp 401–404

Wills RBH, Balmer N, Greenfield H (1980) Composition of Australian foods 2. Methods of analysis. Food Technol Aust 32:198–204

Wills RBH, McGlasson WB, Graham D, Lee TH, Hall EG (1981) Postharvest: an introduction to the physiology and handling of fruit and vegetables. NSW Unieversity Press, Kensington, Australia

Wills RBH, Scriven FM, Greenfield H (1983) Nutrient composition of stonefruit (*Prunus* spp.) cultivars: apricot, cherry, nectarine, peach, plum. J Sci Food Agric 34:1383–1389

Wills RBH, Lim JSK, Greenfield H (1987) Composition of Australian foods. 40. Temperate fruits. Food Technol Aust 39:520–521

Woodroof JG, Luh BS (1975) Commercial fruit processing. AVI Publishing, Westport, CT, pp 421–415

Almond Nut Analysis

A. Vezvaei and J.F. Jackson

1 Introduction

A considerable amount of data on the nutritional aspects of almond nuts is available in the scientific literature, including quantitative analysis for such components as oil, protein, sugars and crude fibre (Kester and Asay 1975; Kester et al. 1990). Other components that have been determined include soluble sugars, amino acids, minerals, gums and mucilage, pectin and cyanide. Some of these substances play an important role in health; for example, there is a negative correlation between usage of dietary fibre and some diseases (Mod et al. 1981). According to Hartwell (1982) the seed and/or its oil are anti-cancer agents, the kernels are valued in diet, for peptic ulcers. It is no surprise that the seeds and/or oil which contain amygdalin or benzaldehyde are widely acclaimed as folk cancer remedies, for all sorts of cancers and tumours, calluses, condylomata and corns. Almond oil is also used for cosmetic, pharmaceutical and pastry production (Vaughan 1970; Rugraff et al. 1982). Bitter almond is used for flavouring extracts and for manufacture of prussic acid (Woodroof 1967). Bitter almonds when eaten in small quantity sometimes produce nettle-rash. When taken in large quantity, they may cause poisoning. The hull contains about 25% sugar and can be used for livestock feed (Weir 1951). In this chapter methods for analysis of some of these compounds are described.

1.1 Determination of Oil Content and Fatty Acid in Almond Kernels

1.2 Introduction

In almond kernels, oil is the major component of the fruit, being 50–65% of dry weight (Souty et al. 1971; Mehran and Filsoof 1974; Filsoof et al. 1976; Nassar et al. 1977; Riquelme 1982; Canellas 1986). The main fatty acids reported are oleic (59–78%) and linoleic (19–30%) together with small amounts of palmitic, palmitoleic, stearic and linolenic acid (Garcia-Olmedo and Marcos-Garcia 1971; Nassar et al. 1977; Riquelme 1982; Canellas 1986). Synthesis of the lipids begins from 50 days after fruit set and increases to the maximum rate at 70 days after fruit set (Munshi et al. 1982). This phenomenon begins when the jelly-like interior of the nuts hardens to a solid material. During kernel development, the oil content

Modern Methods of Plant Analysis, Vol. 18
Fruit Analysis
Edited by H.-F. Linskens and J.F. Jackson

displays a double sigmoid pattern. Oleic and linoleic acids accumulate to similar levels, but the linolenic acid content decreases during development (Munshi and Sukhija 1984). The oil fraction during the ripening varies from 35 to 61% at harvest time according to the cultivar examined. In almond, as well as in other oil seeds, phospholipids and glycolipids are first synthesized. Later, at an intermediate stage, begins the accumulation of the acylglycerides, a fraction that represents about 95% of the oil in mature fruits (Shing and Privett 1970; Munshi et al. 1982).

1.3 Total Oil Determination (Ether Extraction)

Oil determination can be measured directly or indirectly (Faichney and White 1985).

1.3.1 Procedure of Direct Method

A Soxhlet apparatus is used for oil extraction. One g of crushed almond kernel is weighed by difference into the clean and numbered paper thimble (Whatman 30 × 10 cm). The thimble with the sample is placed in a Soxhlet extraction unit. a dried and tared 250-ml round-bottomed flask (with several glass beads) is placed under the extraction unit. Solvent Shell ×55 (150 ml) is added to the flask and the flask heated by an electric plate. The volatilized solvent is condensed by means of a water condenser during extraction. After extraction, the Soxhlet unit is removed and the solvent remaining in the flask collected by distillation. To avoid oxidation of the oil, a small amount of solvent is added before distillation. The flask, beads, and oil are carefully dried in an oven at 105 °C overnight, cooled in a desiccator and weighed. The difference between the weight of the flask before and after gives the amount of the oil in the sample.

1.3.2 Procedure of Indirect Method

One g of crushed almond is folded into filter paper (Whatman No. 541), carefully, to contain sample, and secured with two paper clips. The packets are numbered with pencil and samples placed into the oven at 105 °C overnight, weighed direct from the oven and placed in Soxhlet apparatus and extracted for 8–16 n. Several packets might be placed in each unit. After extraction, the packets are removed from the Soxhlet apparatus and allowed to dry in an oven at 105 °C for at least 6 n. Samples are weighed direct from the oven after 20 min cooling in the desiccator. Crude fat is the loss of the weight and is expressed as a proportion of the sample dry matter determined simultaneously. The amount of oil generally found in almond is between 45 and 55%.

1.4 Determination of Oil Content and Fatty Acid (Lourdes et al. 1988)

1.4.1 Extraction

First, samples are dried and homogenized. Oil is extracted with diethyl ether in a Soxhlet extractor for 18 h. The extract must be filtered, water eliminated with Na_2SO_4 and solvent removed.

1.4.2 Gas Chromatography Analysis

Each 100 mg of oil is treated with 6 ml of esterification reagent prepared by mixing 57.5 ml of benzene with 172.5 ml of methanol and 2 ml of concentrated H_2SO_4. Fatty acid methyl esters are then extracted with hexane. Analysis is carried out by GLC, and examined under isothermal conditions. The column, injector and detector temperature used are 190, 240 and 240 °C respectively. Nitrogen flow is 30 ml/min with 0.2 μl of sample injected. Identification and quantification of fatty acid is performed by comparing the retention data and the areas under the peaks of standards with samples. The "dead" volume is determined by the Peterson and Hirsch method as proposed by Grobler and Balizs (1974). For standard determination a homologus series of *n*-alkanes (C_{10}–C_{24}) can be injected.

2 Determination of Total Nitrogen (Crude), Protein, Amino Acid and Free Amino Acid

2.1 Determination of Total Nitrogen (Crude) with the Kjeldahl Method

The principle involved with this method is that the nitrogen of protein and other compounds are transferred into ammonium sulphate by acid digestion with boiling sulphuric acid. The acid digest is cooled, diluted with water, and made strongly basic with sodium hydroxide. The ammonia is released and distilled into the boric acid solution. The ammonia in the boric acid solution is titrated with standardized sulphuric acid solution (Scales and Harrison 1920).

2.1.1 Extraction and Analysis

After crushing almond kernels, 1 g is weighed on a filter paper, the paper folded and twisted around the sample and introduced into the Kjeldahl test tube. Run two reagent blanks (with filter paper) through all stages of procedure and subtract blank titration from the sample titration. Two blanks are sufficient for a day's run.

Add special Kjeldahl catalyst tablet (each tablet contains: 1.0 g sodium sulphate, anhydrous and the equivalent of 0.01 g of selenium). Position the rake of the tube inside the fume hood and add 15 ml of concentrated sulphuric acid with an automatic dispenser. Samples are heated at 400 °C in the 1015 Tecator digester under the fume hood. During the first 10 min of digestion, the tap water is full on for cooling the samples, after which time, turn the water down for a further 40 min. Remove the test-tube rack from the digester. The solution should be clear and all the carbon oxidized. Add 50 ml distilled water per tube. For total nitrogen content an Auto Kjeltec 1030 Analyser (Tecator) is used, which automatically measures ammonia released after alkali addition. The prepared tubes are placed in pre-adjusted Kjeltec Auto 1030 Analyser for 3 min and nitrogen content of the samples determined. Crude protein content can be estimated from this value. Protein content varied between 16 to 21% of dry weight in the almond kernel. For determination of moisture content, another 1 g of sample is weighed in a glass dish and put into an oven at 100 °C for 24 h. The sample is cooled in a desiccator and weighed again.

2.2 Determination of Protein and Amino Acid

Dried and defatted samples are obtained for extraction of crushed almond with 0.5 N NaOH (Saura-Calixto et al. 1982). Protein content is determined by the method of Gornell-Biuret (Gornall et al. 1949) or the Lowry (Lowry et al. 1951) spectrophotometric method. Egg albumin can be used as protein standard in both cases. Hydrolysis of protein to amino acid is carried out on 50–60 mg of defatted sample by heating with 6 N HCl containing 4% thioglycollic acid, for 21 h at 110 °C in an atmosphere of nitrogen. Amino acids are determined with a Beckman automatic analyzer using column of AA-20 resin. Buffer solutions of pH 3.25, 4.12 and 6.40 at 50 °C are used for elution. Sample size is 50 µl. The ninhydrin reaction is used to estimate each amino acid: the wavelength is set at 570 nm for all amino acids, except for proline, which is determined at 440 nm.

2.3 Determination of Free Amino Acid

The free amino acid content of crushed almond can be determined by extraction with 150 ml of a methanol-chloroform-water (69:25:15) mixture (Young 1974). After the solvent is removed, samples are dissolved with a pH 2.2 buffer and analyzed with an automatic analyzer under the same conditions as described above.

3 Determination of Carbohydrate in Almond Kernels

3.1 Ethanol Soluble and Other Non-Structural Carbohydrate in Almond Kernels

This method has been applied by Dubois et al. (1956) for the determination of simple sugars, oligo-saccharides, poly-saccharides and methyl ethers with free or potentially free reducing groups. These compounds give and orange-yellow colour when treated with phenol and concentrated sulphuric acid. The reaction is sensitive and the colour stable.

3.1.1 Extraction

Grind almond kernels finely using a mortar and pestle, add 1 g of sample to boiling 80% ethanol and continue to boil for a further 15 min. Decant slurry into a 10- or 50-ml centrifuge tube, and wash mortar and pestle with 5 ml of 80% ethanol and add to decanted slurry. Centrifuge for 10 min at 2000 rpm to separate solids and supernatant. Transfer the supernatant with a pipette into a 25-ml volumetric flask. Add 5 ml of 80% ethanol to the pellet and heat the slurry to boiling in the water bath for 5 min (prevent frothing of the samples with a glass rod), transfer the supernatent with the pipette to the 25-ml volumetric flask and make up to volume with 80% ethanol.

3.1.2 Analysis

Reagents used include 500 ml concentration sulphuric acid in an automatic dispenser (5 ml volume), 500 ml 5% phenol solution and standard sucrose solution (10 mg/100 ml). Select a series of 15 × 18 mm glass test tubes which are clean and free from dust or cellulose fibres. In the first step take 0.1 ml of 80% ethanol extract and add 0.9 ml of water, simultaneously adding 1 ml of 5% phenol. Pump a jet of 5 ml of concentrated sulphuric acid to the central part of the sample and mix (Careful: this step is potentially dangerous, and solution comes to the boil.) A face mask must be worn, and acid must be added behind the screen in the fume hood. Measure the optical density at 490 nm when the tubes have cooled, and standardize the procedure with standard of sucrose (5–70 μg per tube). The amount of ethanol soluble sugar in kernels of Nonpareil almond is between 2.5 and 3.5 mg/g.

3.2 Determination of Sugars and Sugar-Alcohols in Almond

3.2.1 Extraction

Kernels are ground so as to pass through a 0.5-mm sieve. Eighty % aqueous ethanol at 50 °C is employed to extract sugars and the solvent removed by vacuum distillation on a Rotavapor apparatus to yield dry residues. Kernel oil had previously been extracted with ethyl ether by using Soxhlet residues. The procedure of Sweeley et al. (1963) is followed to prepare trimethylsityl derivatives (Me_3Si derivatives). One ml of anhydrous pyridine, 0.6 ml of hexamethyldisilazane and 0.4 ml of trimethylchlorosilane are added to dry samples containing 10–20 mg of carbohydrates, and shaken occasionally. Derivatization occurs at room temperature with quantitative yields. Similar treatment is carried out with 10 mg of each standard sugars equilibrated for 24 h in an aqueous solution, which is then evaporated to dryness at 40 °C under vacuum before trimethylsilylation (Sawardeker and Sloneker 1965).

3.2.2 Gas Chromatography Analysis

Assays are performed under isothermal conditions. The operating conditions for Me_3Si derivatives of samples and standard are as follows:

Flow of the carrier gas (nitrogen), 35 ml/min; column temperature 170, 210, 230 and 290 °C. Detector temperature (flame ionization), 300 °C, sample size, 0.3 μl. In order to determine Kovats' indices (I) of Me_3Si derivatives, several mixtures of *n*-alkaline (C_{10}-C_{24}) are injected under the same operating conditions. The Kovats' indices are a means of calibrating the columns. The gas holdup time is calculated by the Peterson and Hirsch method (1959), and (I) values are statistically calculated by the computer methods of Grobler and Balisz (1974).

Identification of carbohydrates are made by comparing the Kovats' indices of samples and standards. Quantitative analysis is determined by comparing the corrected areas under each peak. Several mixtures of standards with known composition are prepared to determine the response factor of Me_3Si derivatives for each sugar and sugar-alcohol at different temperatures. Response factor are used to correct peak areas, and quantitative analysis is made by the internal normalization method (Tranchant 1978). However, it is well known that the precision of Kovates' indices is greater than that of chromatography hold up times and the results obtained are more acceptable and uniform. Assays between 170 and 210 °C are adequate for the separation of Me_3Si mono-saccharides and sugar alcohols. Higher temperatures are necessary to obtain adequate resolution of oligosaccharides, e.g. 230 °C for di-saccharides and 290 °C for raffinose. The amounts of total carbohydrate and ethanol soluble sugars in almond kernels vary from 17.3 to 21.5 and 3 to 8% respectively.

4. Determination of Gums and Mucilage

4.1 Introduction

Plant gums are defined as those substances of plant origin which are obtained as exudations from the fruit, trunks or branches of trees spontaneously or after mechanical injury of the plant by incision of the bark, by removal of branch, or after invasion by bacterial or fungi. It has also been suggested that gums result from normal plant metabolism (Jones and Smith 1949). The gums appear to be produced by some protective mechanism in order to seal off the injured portion from attack by organisms. Almond, like other members of the Rosaceae, including the *Prunus* genus, contains gums and mucilage which are complex carbohydrates (Ryugo and Labavitch 1978). These substances often appear on the surface of tissues with injuries caused by physical or insect and bacterial diseases. Almond infection by hull rot organisms often induces gummy fruits, sometimes at the junction of the pedicel, thus preventing fruit abscission at harvest time (Browne et al. 1977).

4.1.1 Sample Extraction and Reducing Sugar Analysis

Immature hulls are separated from endocarp and frozen. Equal amounts of frozen hull and water are homogenized with a blender and filtered through several layers of cheesecloth. The filtered liquid is centrifuged at 750 *g* and the supernatant is decanted and stored in frozen state until analysis. A 50-ml aliquot of supernatant is acidified with 1 ml of 6 N HCL. Add 450 ml of 95% ethanol slowly. The mixture is held overnight and then centrifuged twice, the precipitate being washed with fresh ethanol between centrifugations. Precipitate is prepared from two to three aliquots of each supernatant sample. The precipitate is then transferred to a flask and refluxed with 40 ml of aqueous 2.5% sulphuric acid for 4 h. On cooling, the suspension is filtered and the filtrate brought to 100 ml with water. Sugar concentration in 1-ml aliquots is estimated with anthrone chromatography paper and Fehling's solution (Wildman and Hensen 1940).

4.1.2 Analysis by Gas Chromatography

Neutral sugar composition of the alcohol insoluble gums and mucilage can be determined according to the method of Albersheim et al. (1967). Glycoside content is determined by gas chromatography of premethylated alditol acetates by the method of Hakomori (1964). Premethylated poly-saccharides are further derivatized and analyzed according to Talmadge et al. (1973), except that the partially methylated aldoses are reduced with sodium borodeuteride rather than sodium borohydride to aid in mass spectral analysis of otherwise symmetrical derivatives. Identification of partially methylated aldital acetates is based on co-chromatography with authentic standards and mass spectral analysis.

4.1.3 Determination of Uronic Acid Composition

Uronic acid compositions is determined colorimetrically using the *m*-hydroxydiphenyl assay described by Bramenkrantz and Asboe-Hansen (1973).

Five mg of gum is partially dissolved by suspending in 2 ml of water and subjecting suspension to a 15-h ultrasonic treatment with a Branasonic 12 ultrasonic probe. Add concentrated H_2SO_4 to the sonicated samples to dissolve completely; 0.2 ml of sample is treated with 1.2 ml of sulphuric acid/tetraborate (0.0125 M solution of tetraborate in concentrated sulphuric acid). The tubes are cooled in crushed ice, and after mixing, heated in a water bath at 100 °C for 5 min. After cooling in a water ice bath, add 20 μl of *m*-hydroxydiphenyl reagent (0.15% solution of meta-hydroxydiphenyl in 0.5% sodium hydroxide; the reagent solution must be kept in the refrigerator covered with aluminium foil for more than 1 month). The tubes are shaken and absorbance at 520 nm determined within 5 min. Quantitative measurement is based on the comparison of absorbance at 520 nm with galacturonic acid standards.

5 Pectin Determination

Pectin in almond kernels can be determined by extracting ground samples with water (soluble pectins) and then 0.05 M HCL at 100 °C for 90 min (insoluble pectins). Pectins can be determined as uronic acid by the carbazole method, using galacturonic acid as standard (Bitter and Muir 1962).

5.1 Method

After extraction of soluble sugars from the ground samples with 80% ethanol, the residue is treated with water to obtain soluble pectins and then 0.05 M HCL at 100 °C for 90 min to obtain insoluble pectin. One-ml aliquots of each pectin fraction (containing 4 to 40 μg galacturonic acid or equivalent) are layered onto 5 ml of 0.025 M sodium tetraborate in sulphuric acid (1.84 specific gravity) in a test tube placed in a rack. The tetraborate-sulphuric acid solution should be cooled to 4 °C before attempting to layer test sample. The tubes are closed with ground glass (or Teflon) stoppers and the rack shaken gently to begin mixing, and finally shaken vigorously with constant cooling to complete mixing. Keep mixture below room temperature. Heat tubes for 10 min in a boiling water bath, cool to room temperature. Add 0.2 ml of 0.125% carbazole in absolute ethanol to each tube, shake to mix, heat a further 15 min in a boiling water bath, cool to room temperature. Read optical density at 530 nm in a colorimeter; sulphuric acid is used as reference optical density.

The method can be automated as described by Berlepsch (1969) or Faichney and White (1985). Almond kernels contain only small amounts of soluble pectin (approximately 110 mg/100 g kernel), but do contain relatively large quantitives of insoluble pectin, between 1 and 2% by weight (Saura-Calixto et al. 1983).

6 Determination of Dietary Fibre in Almond

6.1 Introduction

For several years now, dietary fibre has received considerable attention because of possible benefits to human health. Fibre is the collective name for the insoluble indigestible cell wall constituents of plant materials; it consists of cellulose (a glucose polymer), hemicellulose (a heterogeneous group with pentosans predominant-hexosans, and poly uronides), lignin (an aromatic compound), protopectin and some nitrogen-containing materials (Van Soest 1963; Worth 1967; Norman 1973; Hellendoorn et al. 1975).

6.2 Methods for Determination of Dietary Fibre

Several methods have previously been developed for the determination of crude fibre, unavailable carbohydrates or dietary fibre. A number of gravimetric, colorimetric, and chromatographic methods have been developed. The Weende method determines only crude fibre and was adopted by the Association of Official Analytical Chemists (AOAC) as early as 1887. In the period from 1963 to 1981, the acid detergent fibre (ADF) and the neutral detergent fibre (NDF) methods for determination of hemicellulose, cellulose and lignin were developed and improved. In 1969, Southgate introduced an enzymatic method for determination of soluble and insoluble dietary fibre, and Hellendoorn et al. (1975) described the enzyme gravimetric method for insoluble dietary fibre. A gas chromatographic (GC) method for determination of soluble and insoluble dietary fibre was developed as a method with stepwise acid hydrolysis followed by GC and/or colourimetric quantification of total, insoluble, and soluble non-starch polysaccharides (Englyst and Cummings 1988). In 1983, Aspinal et al. reported a gravimetric method for determination of total insoluble dietary fibre. The determination of total dietary fibre was given official final sanction by AOAC. In 1988, Prosky et al. presented an enzymatic gravimetric method for determination of insoluble, soluble and total dietary fibre in food based on the method of Aspinal et al. (1983).

6.3 Neutral and Acid Detergent Fibre, Cellulose and Lignin

Almond hulls used as a food supplement in animal feeds have a relatively high fibre content showing 10–15% neutral detergent fibre, acid detergent fibre 20–35%, crude fibre 10–25%, cellulose 20–35% and crude lignin 7–15% (Homedes et al. 1993). Neutral detergent fibre (NDF) can be determined by treating ground sample with 200 ml neutral detergent solution (sodium lauryl sulphate pH 7.0) for 1 h at boiling and drying residue, which gives NDF (Van Soest 1963). Washing with water and acetone, then drying, gives hemicellulose (by weight difference). If kernels or hulls are treated with acid detergent solution (cetyl methyl ammonium bromide in 0.5 M sulphuric acid) for 1 h at boiling, wash with water and acetone; drying gives acid detergent fibre (ADF) (Van Soest and Robertson 1979). If the sample is treated with $KMNO_4$ and 72% sulphuric acid, lignin and cellulose are solubilized, allowing determination of these components as well.

6.4 Sequential Enzymatic Method for Fibre

This method evolved from those of Furda (1981), Aspinal et al. (1983) and Schweizer (1989) and, basically, is described by Prosky et al. (1988). The general principle of this AOAC-approved method is the quantitative removal of small molecular weight components, protein and starch from plant or food materials using enzymatic hydrolysis and extraction. The sequentional enzymatic hydrolysis is by alpha-amylase, followed by protease and amyloglucosidase. This not only hydrolyzes starch and proteins, but after starch is solubilized by alpha-amylase, the plant matrix becomes more open and accessible to the protease and amyloglucosidase that complete the hydrolysis, which is followed by extraction purification. The isolated dried residue is determined gravimetrically and the residual protein and ash are determined and subtracted. The method for total dietary fibre (TDF) is modified to give separate determination of soluble fibre(SF) and insoluble fibre (IF).

Half to 1 g sample is incubated at 100 °C with 50 ml buffer pH 6 and 50-100 μl alpha-amylase (approx.1 mg) and allowed to cool. The pH is adjusted to 7.5 with 10 ml 0.171 N sodium hydroxide and the sample incubated with 5 mg protease for 30 min at 60 °C to hydrolyze the protein, cooling the sample. After the pH is adjusted to 4.5 (10 ml 0.205 M H_3PO_4) and incubated with 200–300 μl amyloglucosidase (approx. 3 mg) for 30 min at 60 °C to hydrolyze the starch dextrins. From this point, the methods for total dietary fibre and soluble/insoluble dietary fibre proceed differently. For determination of total dietary fibre, the soluble portion is precipitated overnight with 280 ml 95% ethanol (4 vol.) at room temperature. After filtration, the residue is washed successively with ethanol and acetone, dried and weighed. A correction is made for ash and protein and amount of total dietary fibre is calculated. For separate determination of soluble and insoluble fibre, the samples are filtered and the residue washed successively with ethanol and acetone, dried and weighed. A correction is made for ash and protein and the amount of insoluble dietary fibre is calculated. The residue is dried and weighed. A correction is made for ash and protein and the amount of soluble dietary fibre is calculated. This method has the limitation that ethanol-soluble polysaccharides such as polydextrose are not recovered, nor are ethanol-soluble indigestible oligo-saccharides such as raffinose or stachyose. The total dietary fibre sometimes contains a fraction of starch that is not hydrolyzed by alpha-amylase or amyoglucosidase. However, this part of the starch may not be enzymatically hydrolyzed in vitro, and should not therefore be considered as part of dietary fibre.

7 Mineral Elements Measurement in Almond Kernels

There are several analytical techniques available for determination of minerals in nuts. For example there is atomic absorption, X-ray fluorescence, flame emission and arc/spark emission spectroscopy, spectrophotometry and specific ion electrodes. With inductively coupled plasma spectrometery (ICPS), it is possible to determine P, K, S, Ca, Mg, Na, Al, Zn, Mn, Fe, Co, Mo and B in plant materials. ICPS has the potential to determine all the nutritional elements (except nitrogen) with a polychromator sequentially scanning monochromator in the single digestion (Zarcinas et al. 1987).

The advantage of this method is that it is simple, quick and precise; plant material is made ready for this method by digestion using nitric acid to allow the simultaneous determination of at least 14 elements. The disadvantage is that it is a relatively expensive method.

7.1 Sample Preparation and Analysis

Almond samples are dried overnight at 65–70 °C before digestion. Almond kernel is ground finely with a mortar and pestle. After weighing 1 g of sample, transfer into a tube with a mark to indicate 25 ml. Add 10 ml of nitric acid (Univar A.R. 70% w/w) with the dispenser under the fume hood. (lab coat, plastic gloves and face mask must be worn for safety reasons at this stage). After adding nitric acid, cover the sample loosely with plastic film and leave overnight or longer in the fume hood. The tubes are then heated for 4 h at 120 °C and increase the temperature to 140 °C. Digestion is continued at this temperature until about 1 ml of acid remains (always the combination of pre-digest vortex and cold start prevents sample from frothing over). After cooling, dilute to 20 ml with 1% v/v nitric acid. Decant into a polystyrene tube (N.B. avoid the transfer of solid). The digest should be stood overnight before ICPS analysis to allow precipitation of any suspended solids. A suitable instrument of use is a 3580B ICP Spectrometer Sim/Seq. and 21 simultaneous channels manufactured by Applied Research Laboratories (Switzerland). Typical amounts of five of these minerals in kernels of Nonpareil cultivar of almond (mg/g) are: K, 11.3 $\pm$ 0.19; P, 7.5 $\pm$ 0.07; Mg, 3.93 $\pm$ 0.04; Ca, 3.60 $\pm$ 0.16; and S, 1.95 $\pm$ 0.04.

8 Determination of Dry Matter and Inorganic Ash

Weigh accurately less than 2 g of sample into a weighed silica crucible, place in oven at 105 °C for 24 h, cool in desiccator and weigh. Calculate dry matter fraction. Place crucibles at random in furnace chamber and switch on furnace, if not starting from cold, hold furnace at about 250 °C until no further smoke is given off.

Increase temperature after 3 h until 550 °C is reached. Cool in the desiccator and weigh. Calculate total ash as a proportion of the sample dry matter.

8.1 Insoluble and Soluble Ash

Weigh accurately no more than 2 g of ground sample into 250-ml beaker, then add 50 ml water at about 39 °C for 1 h with shaking at 100 cycles per min. Filter through a Whatman No. 541 ashless filter paper and wash three times with 50 ml water. Fold paper to enclose extracted sample, place in weighed silica crucible and dry in oven overnight. Place crucible at random in furnace and proceed as for total ash. Calculated insoluble ash as a proportion of the sample dry matter and soluble ash by difference from total ash (above). In mature almond kernels the moisture content is generally between 5 and 6% dry matter and ash between 3 and 4% dry matter.

9 Qualitative and Quantitative Determination of Cyanide in Almond

9.1 Qualitative Method

Sodium picrate soaked paper is used for qualitative cyanide determination in almond kernels. It is an easy, quick method for assay of a large number of samples. Sodium picrate paper is used as explained by (Guignard 1906; Kingsbury 1964; Horwitz 1980; Brinker and Seigler 1992). Almond kernel is ground in a cold mortar and pestle with liquid nitrogen. One g of this sample is placed into a 10-ml plastic test tube and add 0.5 ml of phosphate buffer pH 6.8 including 20 μl beta-glucosidase (100 μg beta-glucosidase in 2 ml of phosphate buffer), and then a wet strip of sodium picrate paper (3 × 1 cm) is suspended over the test tube, the lid clamped shut and sample incubated for at least 4 h at 37 °C. Where cyanide is present in the kernels, the sodium picrate paper changes from yellow to orange, or to deep orange at higher cyanide concentrations. We have examined commercially grown kernels from almond cultivars Nonpareil, Price, Carmel, Peerless, Ne-Plus Ultra, Keane's Seedling, Fritz, Mission and a wild almond species *Prunus lycioides* (Spach.) Schneider. Kernel samples from the wild species showed a deep orange colour, indicating a relatively high level of cyanide. For Fritz and Mission kernels, sodium picrate paper changed to a pale orange colour, the colour density in Mission being lower than Fritz, indicating a low and harmless level of cyanide in both these cultivars. The other cultivars showed no detectible cyanide by this method.

9.2 Quantitative Method

In principle, quantitative analysis of cyanide in plant tissues involves hydrolysis of the cyanogenic glucoside and liberation of HCN which is trapped in basic solution and determined colorimetrically (Nahrstedt et al. 1981; Brinker and Seigler 1989).

The amount of cyanide in the kernel can be estimated by preliminary qualitative test (sodium picrate paper see above). Each Kernel is ground separately in a cold mortar and pestle with liquid nitrogen. To a 100-ml flask, inside which a glass vial forms the central part of the bottom of the flask, add 1 g of ground sample outside the central vial or well. Then 0.5–0.6 ml of a solution of sodium phosphate pH 6.8 including 20 μl beta-glucosidase (100 μg beta-glucosidase in 2 ml of phosphate buffer) is added to the ground sample in that part of the flask outside the central vial in the bottom of the flask. Inside the central vial is placed 1 ml of freshly made up 1 M sodium hydroxide. The flask is sealed and incubated for 4 h at 37 °C. During incubation, the HCN released from sample is trapped by sodium hydroxide as NaCN inside the vial. Cyanide is determined according the Lambert method (Lambert et al. 1975). In this method, 0.1-aliquots of the sodium hydroxide are removed from samples and put in a set of 150 × 18 mm test tubes and diluted to 1 ml. Add 0.5 ml 1 M acetic acid, 5 ml of succinimide/*N*-chlorosuccinimide (2.5 g succinimide and 0.25 g *N*-chlorosuccinimidein 1 l of water) and 1 ml of barbituric acid/pyridine reagent to each tube. The latter reagent is made up by adding 12 g barbituric acid is mixed with a small amount of water to make a paste, then 120 ml pyridine is added with enough water to give a final volume of 400 ml. The tubes must be shaken or vortexed vigorously and allowed to settle for 10 min. In the presence of the cyanide a purple colour appears and the optical density measured at 580 nm. The colour is stable for at least for 30 min. For preparation of a standard curve, aliquots of a standard NaCN solution (Brinker and Seigler 1989) are made up to 1 ml with 0.1 sodium hydroxide.

A standardized solution of NaCN is prepared as follows: a 0.02 M solution of NaCN in 0.1 N NaOH is prepared (0.980 g NaCH/l) and the exact amount of CN^- in the solution is determined by a modified Liebig titration method (Kolthoff and Sandell 1952; Ayres 1958; Vogel 1961), as follows: (1) Prepare a standard 0.1 M $AgNO_3$ solution: weigh out 16.987 g of dried $AgNO_3$ (as carefully as possible) and dilute to 1 l with water in a volumetric flask. This solution should be stored in the dark. (2) Pipette 25 ml of the NaCN solution to a 50-ml flask equipped with a stirring blade (do not pipette by mouth). (3) Add 1.5 ml 6 N NH_4OH and 0.5 ml 10% KI. (4) Titrate (with stirring) with the standard $AgNO_3$ solution (use a 10-ml burette) until the solution remains faintly opalescent or turbid; this endpoint is easier to see against a black background (about 2.4 ml $AgNO_3$ solution should be required). (5) Calculate the concentration of CN^- in the solution. Multiply the number of ml of $AgNO_3$ necessary to reach the endpoint by 0.005204 g CN^-/ml 0.1 M $AgNo_3$ to determine the amount of CN^-in the solution (g/ml). Aliquots (1 ml) of the NaCN solution (which should be close to 500 μg CN^-/ml) should then be diluted in a volumetric flask with 0.1 N NaOH to 100 ml to give a solution with a final concentration of approximately 5 μg CN^-/ml.

The amount of cyanide determined in the kernels of some almond cultivars is as follows: Non-pareil (0.016 ± 0.002 μg/g), Mission (0.031 ± 0.003 μg/g), Fritz (0.072 ± 0.006 μg/g), these amounts being extremely low and harmless, and bitter almond (0.421 ± 0.028 μg/g).

References

Albersheim P, Nevins DJ, English PD, Karr A (1967) A method for the analysis of sugars in plant cell wall poly-saccharides by gas-liquid chromatography. Carbohydr Res 5:340-345

Aspinal NG, Johansson CG, Halmer H, Siljestrom M (1983) Rapid enzymatic assy of insoluble and soluble dietary fiber. J Agric Food Chem 31:476-482

Ayres GH (1958) Quantitative chemical analysis. Harper, New York, 392 pp

Berlepsch K von (1969) Two automated procedures for Determination of uronic acid in acid glycosaminoglycans. Analyt Biochem 27:424-435

Bitter T, Muir HA (1969) A modified uronic acid carbazole reaction. Anal Biochem 4:330-334

Bramenkrantz N, Asboe-Hansen G (1973) New method for quantitative determination of uronic acids. Anal Biochem 54:484-489

Brinker AM, Seigler DC (1989) Methods for detection and quantitative determination of cyanide in plant materials. Phytochem Bull 21:24-31

Brinker AM, Seigler DC (1992) Determination of cyanide and cyanogenic glycosides from plant In: Linskens HF, Jackson JF (eds) Modern methods of plant analysis new series, vol 13. Springer, Berlin Heidelberg New York

Browne LT, Ogawa JM, Gashaira B (1977) Search continues for control of almond hull rot. Calif Agric 31:16-17

Canellas J (1986) Estudio de Componentes Quimicos de Variedades de Almendra de Baleares. Doctoral Thesis, Univiersity of Illes Balears, Palma de Mallorca, spain

Dubois M, Gilles KA, Hamilton JK, Rebers PA, Smith F (1956) Colorimetric method for determination of sugars and substances. Anal Chem 28:350-356

Englyst HN, Cummings JH (1988) Improved method for measurement of dietary fibre as non-starch polysaccharides in plant foods. J Soc Off Anal Chem 71(4):808-814

Faichney GJ, White GA (1985) Methods for the analysis of feeds eaten by ruminatnts. Division of Animal Production, Ian Clunies Ross Animal Research laboratory, PO Box 239, Blacktown, NSW 2184, Australia

Filsoof M, Mehram M, Farrohi F (1976) Determination and Comparison of oil characteristics in Iranian almond, apricot and peach nuts. Fetter Seifen Anstrichm 78:150-151

Furda I (1981) Enzymatic fiber determination. In: James WPT, Teander O (eds) The analysis of dietary fiber in food. Marcel Dekker, New York

Garcia-Olmedo R, Marcos-Garcia MA (1971) Contribucion al Estudio de los Aceites de Frutos Secos Espanoles. Composicion Acidiac. Anal Bromatol 23:253-258

Gornall AC, Bardawill CS, David MM (1949) Determination of serum proteins by mean of the Biuret reaction. J Biol Chem 177:751-766

Grobler A, Balizs G (1974) Computer program for a statistical calculation of retention indices. J Chromatogr Sci 12:57-58

Guignard 1 (1906) Le haricot à acide cyanhydrique (*Phaseolus lunatus*). Etude historique, botanique et chimique. Nouveau procédé pour déceler l'acide cyanhydrique. Bull Sci Pharmacol 13:401-419

Hakomori S (1964) A rapid permethylation of glycolipid and poly-saccharide catalyzed by methylsulfinyl carbanion in dimethyl sulfoxide. J Biochem 55:205-208

Hartwell HJ (1982) Plant used aginst cancer. A survey reissued in one volume. Quarterman Publications, Lawrence, MA 710

Hellendoorn EW, Noordhoff MN, Slagman J (1975) Enzymatic determination of the indigestible residue (dietary fibre) content of human food. J Sci Food Agric 26:1461–1468
Homedes JM, Roura E, Keim NL, Brown DL (1993) Almond hulls in Swiss diet reduce diet fat. Calif Agric 47:27–28
Horwitz W (ed) (1980) Official methods of analysis of the association of official analysis chemists, 13th edn. Assoc Off Anal Chem, Washington, DC, 433 pp
Jones JKN, Smith F (1949) Plant gum and mucilage. Adv Carbohydr Chem 4:243–291
Kates M (1872) Method in lipidology. In: Work TS, Work E (eds) Laboratory techniques in biochemistry and molecular biology, vol 3. North Holland, Amsterdam, 267 pp
Kester DE, Assay RA (1975) Almond. In: Janick J, Moore JN (eds) Advance in fruit breeding. Purdue University Press, West Lafayette, IN, pp 387–419
Kester DE, Gradziel TM, Grasselly C (1990) Almond (*Prunus*) In: Moore JN, Ballington JR Jr (eds) Genetic resources of temperature fruit and nut crops. International Society for Horticultural Secience, Wageningen
Kingsbury JM (1964) Poisonous plants of the united state and Canada. Prentice-Hall, Englewood Cliffs, NJ, 25 pp
Kolthoff I, Sandell E (1952) Textbook of quatitative inorganic analysis. Macmillan, New York
Lambert JL, Romasmy J, Raukstelis JV (1975) Stable reagents for the colorimetric determination of cyanide by modified Konong reaction. Anal Chem 47:916–918
Lourdes S, Canellas J, Saura-Calixto F (1988) Oil content and fatty acid composition of developing almond seeds. J Agric Food Chem 36:695–697
Lowry OH, Poseburgh NJ, Farr AA, Randall RJ (1951) Protein measurement with the folin-phenol reagent. J Biol Chen 193:265–275
Mehran M, Filsoof M (1974) Characteristics of Iranian almond oils and nuts. J Am Oil Chem Soc 51:433–435
Mod RR, Ory RL, Morris NM, Normand FL (1981) Chemical properties and interactions of rice hemicellulose with trace minerals in vitro. J Food Chem 29:449–454
Munshi SK, Sukhija SP (1984) Compositional and biosynthesis in developing kernels of almond (*Prunus amygdalus* Batch). J Sci Food Agric 35:689–697
Munshi SK, Sukhija Sp, Bahatia IS (1982) Lipids biosynthesis in developing kernels of almond (*Prunus amygdalus* Batsch). Phytochemistry 22:79–83
Nahrstedt A, Erb M, Zinsmeister H (1981) Methods of liberating and estimating hydrocyanic acid from cyanogenic plant material. In: Vennesland B, Conn EE, Knowles CJ, Westley J, Wissing F (eds) Cyanide in biology. Academic Press, New York, pp 461–471
Nasser AR, E1-Tahawi B S, El-Deen AS (1977) Chromatographic identification of oil and amino acid constituents in kernels of some almond varieties. J Am Oil Chem Soc 54:553–556
Norman AG (1973) The biochemistry of cellulose, the polyuronides, lignin. Clarendon Press, Oxford
Peterson ML, Hirsch J (1959) A calculation for locating the carrier gas front of a gas-liquid chromatogram. J Lipid Res 1:132
Prosky L, Aspinal NG, Schweizer TF, DeVries JW, Furda I (1988) Determination of insoluble, soluble, and total dietary fibre in foods and food products: interlaboratory study. J Assoc Off Anal Chem 71(5):1017–1023
Riquelme F (1982) Caracteristicas de las Almendras Murcianas y su Estabilided. Doctoral Thesis, Murcia, Spain
Rugraff L, Demanze Ch, Karleskind A (1982) Etude des Huiles Oléiques à Usage Cosmetique et Pharmaceutique. Parfums Cosmet Aromes 43:59–67
Ryugo K, Labavitch J (1978) Gum and mucilage in hull of almonds. J Am Soc Hortic Sci 103:568–570
Saura-Calixto F, Jaime C, Lourdes S (1982) Dietary fiber and components of the nitrogen-free extract of almond kernels. J Sci Food Agric 34:1419–1422
SawardekerJS, Sloneker JH (1965) Quantitative determination of monosaccharides by gas-liquid chromatography. Anal Chem 37:945–947
Scales FM, Harrison AP (1920) Boric acid modification of the Kjeldahl methods for crop and soil analysis. J Ind Eng Chem 12:350–352
Schweizer TF (1989) Dietary fiber analysis. Lebensm Wiss Technol 22:45–59

Shing H, Privett OS (1970) Studies on the glysolipids and phospholipids of immature soybeans. Lipids 5:692–695

Souty M, Andre PBJ, Jacquemin G (1971) Etude sur la Qualité des amandes *Amygdalus communis* L.: Variabilité de Quelques Caractéres Biochimiques. Ann Technol Agric 20:121–130

Sweeley CC, Bentley R, Makita M, Wells WW (1963) Gas liquid chromatography of trimethylsilyl derivatives of sugars and related substances. J Am Chem Soc 85:2497–2507

Talmadge KW, Keegstra K, Bauer WD, Albersheim P (1973) The structure of plant cell walls. I. The macromolecular components of suspension-cultured sycamore cells with a detailed analysis of the pectic polysaccharides. Plant Physiol 15:719–725

Tranchant J (1978) Manual practico de comatografia en fase gaseosa. Toray-Masson SA, Barcelona, Spain, pp 241–249

Van Soest PJ (1963) Use of detergents in the analysis of fibrous feeds. II. A rapid method for the determination of fibre and lignin. J Assoc Off Agric Chem 46:829–835

Van Soest PJ, Robertson JB (1979) System of analysis for evaluating fibrous feeds. Proc meeting on standardization of analytical methodology for feeds, Ottawa, August 1979

Vaughan JG (1970) The Structure and utilization of oil seeds. Chapman and Hall, London, pp 49–53

Vogel AI (1961) Textbooks of inorganic analysis-including instrumental analysis. Longman, London

Weir WMC (1951) Almonds hulls and feed. Calif Agric 5(9):3

Wildman SG, Hansen E (1940) Semi-micro techniqe for determining reducing sugars. Plant Physiol 15:719–725

Woodroof JG (1967) Tree nuts: production, processing, products, vol 1. AVI Publishing, New York

Worth HG (1967) The chemistry anf biochemistry of pectic substances. Chem Rev 67:465–473

Young CT (1974) Effect of harvest date and maturity upon free amino acid levels in three varieties of peanuts. J Am Chem Soc 51:269–273

Zarcinas BA, Cartwright B, Spounser LR (1987) Nitric acid digestion and multi-element analysis of plant material by inductively coupled plasma spectrometry. Common Soil Sci Plant Anal 18:131–146

Genetic Characterization of Chestnut

G. Mulcahy Bergamini, A. Trautweiler, H.F. Linskens and D.L. Mulcahy

1 Introduction

Chestnut is an edible fruit of deciduous trees of the genus *Castanea*, belonging to the family of Fagaceae. The genus *Castanea* includes 13 species. They all are natives of the temperate zone in Asia, Europe and North America, where they cover more than 200 million acres.

The fruit of the chestnut provided and still provides an important food for man, used in boiled, baked, roasted, stuffed or confectionery form, and as feed for animals as well. In 19th century Italy, 30 chestnut trees were considered a valuable dowry for the young bride and a warrant for survival in years of scarcity of food. In addition, the other products of the chestnut trees were widely used in former times as fuel (burrs), timber, and leaves as litter for cattle in the stables. Chestnut was, in former times, a valuable source of tannin for the production of leather.

The nuts of chestnut are borne in a spiny wrapper (involucrum), the burr. The fruits of the chestnut are, in botanical terms, nuts, the shell of which is developed from the ovary wall. The edible kernel of the nut is the embryo, mostly consisting of the huge cotyledons, the small root and the embryonic shoot. The embryo is enclosed into a sometimes felty membrane, the pellicle, which develops from the integuments.

2 The Importance of Chestnut Genetic Analysis

The genetic analysis of chestnut has become increasingly important since the rise of the chestnut blight disease at the end of the 19th century, dramatically decreasing the productivity of this species. The chestnut disease is caused by a systemic fungus, callled *Cryptonectria parasitica*, which "keeps smoldering at the roots and sending up new shoots, till another parasite shall come to end the blight" (Robert Frost, Evil Tendencies Cancel, in Latham 1949), which happened recently in the form of a hypervirulent strain of the parasite. The chestnut blight involves also the production of two endotoxins, dissporthin and skyrin, which enter the water transport system of the tree, causing the wilting symptoms characteristic for the disease.

Modern Methods of Plant Analysis, Vol. 18
Fruit Analysis
Edited by H.-F. Linskens and J.F. Jackson

Propagation of the chestnut by seeds and by vegetative propagation by grafting (nursery or nut grafting, stooling) was responsible for the rapid spread of the blight disease. Breeding for disease and insect resistance became essential; insects are instrumental in the diffusion of the diaspores of the highly virulent fungus. Fortunately, certain species of *Castanea*, like the Chinese and the Japanese chestnuts (*Castanea mollissima* Bl. and *C. crenata* Sieb & Zucc., respectively) have some field resistance, whereas the American and the European chestnut (*Castanea sativa* Mill. and *C. dentata* Burkh., respectively) are highly susceptible. Identification of blight-resistant strains at an early stage of development is badly needed for any effort at selecting resistant genoytpes.

Chestnut trees are typically monoecious, but chestnut is self-incompatible, which makes cross-pollination a necessity for crop production. Recognition of the incompatible and compatible strains in an early stage of development is essential for the planting of economically valuable chestnut gardens.

Chestnut fruits/nuts are increasingly wanted on the market because of the low fat content compared with filbert and pecan; they are readily digested in roasted or boiled form, and have a low caloric content. Chestnut confections are highly priced on the market.

Although identification of chestnut samples, strains and provenances can be done using fruits and leaves, the availability of molecular methods would be useful and essential in breeding for vigour, form and increasing nut production. These, however, require samples of purified DNA, and the presence of secondary metabolites in leaves of chestnut species requires elaborate extraction and purification methods.

3 DNA Extraction and Purification

1. Mix 0.19 g sodium bisulfite with 50 ml extraction buffer (see * notes).
2. Grind 1 g of chilled leaf tissue in 6 ml extraction buffer (4 °C) in chilled mortar. Pour into 50-ml conical tube.
3. Rinse mortar with 1.5 ml extraction buffer. Add to conical tube.
4. Add 7.5 ml lysis buffer (see ** notes).
5. Add 2.5 ml of 5% Na N-lauroylsarcosine (w/v). Vortex or shake thoroughly.
6. Incubate at 65 °C for 20 min, vortexing every 5 or 10 min.

3.1 Protein Precipitation

7. Add 5 ml chloroform:phenol:isoamyl alcohol (24:25:1).
8. Add 3 ml 5 M sodium perchlorate.
9. Add 0.5 ml 45% EtOH saturated with SDS.
10. Vortex or hand shake for 5 min.
11. Centrifuge at top speed for 10 min. Remove and save supernatant.
12. Repeat steps 8–12 until supernatant is clear.

13. Add 5 ml chloroform:isoamyl alcohol (24:1) (to eliminate phenols). Vortex, centrifuge on high for 10 min. Save supernatant for step 14.

3.2 DNA Precipitation

14. To supernatant from step 13, add 2/3 vol isopropanol, vortex briefly.
15. Centrifuge at 2000 rpm for 5 min.
16. Wash pellet with 70% EtOH 2×, centrifuging each wash.
17. Vacuum dry pellet.

3.3 DNA Purification

18. Resuspend pellet in 5 ml of pre 2-BE buffer (25 mM boric acid/Tris (pH 7.6) containing 1.25 mM Na_2EDTA and 100 mM NaCl).
19. Add 2 ml of 2-BE (0.4 vol of the 5 ml in step 19), store on ice for 30 min.
20. Centrifuge at 20 000 *g* for 10 min at 0 °C (pellets carbohydrates).
21. Remove and measure supernatant. Add 0.6 vol of 2-BE, store on ice 30 min.
22. Centrifuge at 20 000 *g* for 10 min at 0 °C (pellets DNA).
23. Wash pellet with:
    0.2 M boric acid/Tris, 10 mM Na_2EDTA (pH 7.6): 2-BE (1:1 v/v).
    Centrifuge on high for 5 min.
24. Wash pellet with 70% EtOH containing:
    100 mM K acetate/acetic acid,
    pH 6. Centrifuge on high for 5 min.
25. Wash pellet with absolute EtOH, vacuum dry pellet, resuspend in TE buffer.

Notes:

*DNA Extraction Buffer:	Sorbitol	63.75 g/l
	Tris	12.10 g/l
	EDTA (disodium)	1.68 g/l

Add H_2O to nearly 1 l, pH to 7.5 with concentrated HCl (about 7 ml). Final volume up to 1 l.

**Nuclei Lysis Buffer:
200 ml 1.0 M Tris-HCl (pH to 8)
200 ml 0.25 M EDTA (93.05 g/l)
400 ml 5.0 M NaCl (292.2 g/l)
20 g CTAB = Hexadecyl trimethyl ammonium bromide
200 ml dH_2O

Dissolve CTAB in H_2O first, add Tris, then EDTA, stir overnight, then add NaCl.

The extracted DNA is dissolved in 400 µl TE buffer, letting it sit at room temperature for at least 30 min. Then tap the tube gently to disperse precipitate. If it does not dissolve, incubate it at 60 °C, but for no more than 10 min. Transfer to Eppendorf tube. The dissolved DNA samples can be stored for short term at 4 °C, for longer periods, however, −20 °C is necessary.

4 DNA Amplification by Polymerase Chain Reaction (PCR)

DNA concentration was estimated by electrophoresis in 0.9% agarose, 1x TBE buffer (0.089 M Tris borate, 0.002 M disodium EDTA) and visual comparison with lambda DNA samples of known concentrations. For use in PCR, the DNA was diluted with sterile distilled water to 5–10 ng/µl and 2.5 µl of this added to 10.15 µl of reaction mixture for PCR. The PCR reaction mixture was 1.25 µl 10x buffer (100 mM Tris-HCl pH 8.3, 500 mM KCl, 15 mM $MgCl_2$, 0.1% Difco gelatin); 1.0 µl dNTPs (2.5 mM each dNTP); 0.55 µl 10 mM $MgCl_2$; 0.5 µl 10 µM primer; 0.1 µl Taq (DNA polymerase), (5 U/µl); 2.0 µl DNA (5–10 ng/µl); 6.75 µl H_2O. The total volume, 12.65 µl, was overlain with one drop of mineral oil. The PCR conditions were 94 °C for 1 min, followed by 40 cycles of 94 °C for 1 min, 34 °C for 2 min and 74 °C for 2 min. Some reactions were run on a Barnstead/Thermolyne thermocycler (Model DB 66935) and others on a Perkins/Elmer thermocycler (Model 480).

Random 10 bp primers used were A05 (= Set A, primer 05), A05 = 5′AGGGGTCTTG3′, and All, CAATCGCCGT, both purchased from Operon Technologies, Alameda, California. Nucleotides (dNTPs) were from both Promega (Madison, WI, USA) and Perkin-Elmer Cetus, Norwalk, CT, USA). Taq polymerase was AmpliTaq (DNA polymerase AS), from Perkin-Elmer Cetus. The DNA size marker was either the 100-bp ladder or the 123-bp ladder, both were purchased from Gibco (Gaithersburg, MD, USA).

To determine the appropriate concentration of DNA for the PCR reaction, dilution series of some samples were run by adding 2.5 µl of a DNA extract, diluted with sterile distilled water, to 1/2, 1/4, 1/8 . . . 1/128 of the original concentration, to 10.15 µl of the reaction mixture. Strong and consistent amplifications resulted from 1/4 to 1/16 of the original concentration. Comparison to lambda DNA of known concentrations indicated that this range was 20 to 5 ng DNA/12.65 µl reaction (DNA + reaction mixture). Higher concentrations usually gave no amplifications and lower concentrations gave significantly weaker bands.

The reliability of the PCR (polymerase chain reaction) is dependent on many external factors, such as false-negative results caused by uncontrolled PCR inhibitor, but also by contamination with foreign DNA. Special attention has to be given to the breakdown of the PCR by contamination through the air: the contamination by pollen grains may be the cause for seasonal presence of pollen grains in the laboratory air stream (St. Pierre et al. 1994).

Protocol for the amplification of chestnut DNA: The reaction mixture is vortexed and covered with a drop of mineral oil in order to reduce evaporation during the amplification process. The tubes are put in the aperture of the thermocycler, which is in advance filled with a drop of mineral oil in order to guarantee close contact between the wall of the cycler and the tube.

The programme of the thermocyclers was:

1 cycle:	94 °C for 1 min
40 cycles	94 °C for 1 min
	34 °C for 1 min
	72 °C for 2 min
1 cycle	72 °C for 5 min

Store at 10 °C.

5 DNA Separation of the PCR Products by Gel Electrophoresis

Amplified DNA was separated in 2% agarose, (1% NuSieve agarose, from FMC Bioproducts, Rockland, ME, USA), 1% agarose (Cat. No. A-6013, Sigma

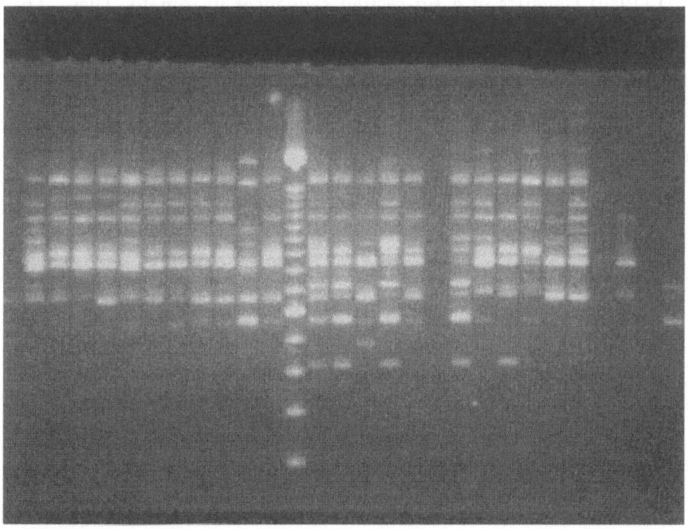

Fig. 1. Sample amplification of *Castanea* spp. DNA, using Operon primer A 18. Details in text

Chem. Co., St, Louis, MO, USA), 1x TBE buffer, and was stained with ethidium bromide.

Results from the two thermal cyclers, Barnstead/Thermolyne thermocycler (48 wells) and Perkin/Elmer thermocycler (48 wells) were interchangeable.

6 Sample of Chestnut Analysis and Discrimination of Strains

Figure 1 illustrates the products of PCR of *Castanea* DNA, Operon primer A18. The centre lane is the Gibco 100-bp ladder, with 600 bp being double the standard intensity. Although many polymorphisms are present, decisive fingerprinting of all individuals would require the use of several different primers. The results, however, demonstrate that the quality of extracted DNA is sufficient to allow PCR analysis.

Acknowledgements. This work was supported by the American Chestnut Foundation by the New York Chapter of the American Chestnut Foundation, and by University of Massachusetts, Office of Biotechnology.

References

Bernatzky R, Tanksley SD (1986) Genetics of actin-related sequences in tomato. Theor Appl Genet 72:314–321

Latham EC (ed) (1949) Complete poems of Robert Frost. Holt, Reinhard, and Winston, New York

Manning K (1991) Isolation of nucleic acids from plants by differential solvent precipitation. Anal Biochem 195:45–50

St Pierre B, Neustock P, Schramm U, Kirchner H, Beln G (1994) Seasonal breakdown of polymerase chain reaction. Lancet 343:673

Subject Index

MIX
Papier aus verantwortungsvollen Quellen
Paper from responsible sources
FSC® C105338

If you have any concerns about our products,
you can contact us on
ProductSafety@springernature.com

In case Publisher is established outside the EU,
the EU authorized representative is:
Springer Nature Customer Service Center GmbH
Europaplatz 3, 69115 Heidelberg, Germany

Printed by Libri Plureos GmbH
in Hamburg, Germany